Atmosphere of Collaboration

This book discusses air pollution in Delhi from scientific, social, and entrepreneurial perspectives. Using key debates and interventions on air pollution, it examines the trajectories of environmental politics in the Delhi region, one of the most polluted areas in the world. It highlights the administrative struggles, public advocacy, and entrepreneurial innovations that have built creative new links between science and urban citizenship. The book describes the atmosphere of collaboration that pervades these otherwise disparate spheres in contemporary Delhi.

Key features:

- Presents an original case study on urban environmentalism from the Global South
- Cuts across science, policy, advocacy, and innovation
- Includes behind-the-scenes discussions, tensions, and experimentations in the Indian air pollution space
- Uses immersive ethnography to study a topical and relevant urban issue

As South Asian and Global South cities confront fast-intensifying environmental risks, this study presents a dialogue between urban political ecology (UPE) and science and technology studies on Delhi's air. The book explores how the governance of air is challenged by scales, jurisdictions, and institutional structures. It also shows how technical experts are bridging disciplinary silos as they engage in advocacy by translating science for public understanding. The book serves as a reminder of the enduring struggles over space, quality of life, and citizenship while pointing to the possibilities for different urban futures being negotiated by variegated agents.

The book will interest scholars and researchers of science and technology studies, urban studies, urban geography, environmental studies, environmental politics, governance, public administration, and sociology, especially in the

Global South context. It will also be useful to practitioners, policymakers, bureaucrats, government bodies, civil society organisations, and those working on air pollution advocacy.

Rohit Negi is Associate Professor of Urban Studies in the School of Global Affairs at Ambedkar University Delhi, India. He has a PhD in geography from the Ohio State University and an MA in urban planning from the University of Illinois at Urbana-Champaign. He is the co-editor of *Space, Planning and Everyday Contestations in Delhi* (2016).

Prerna Srigyan is a PhD Researcher in the Department of Anthropology at the University of California, Irvine, USA. She has an MA in environment and development from Ambedkar University Delhi and a BSc (Hons) in chemistry from Hindu College, University of Delhi. She works on science pedagogy and politics of collaboration, focusing on transnational science and technology studies.

"This is a story-rich, theoretically framed and very practical guide to air pollution politics in Delhi that points to exciting possibilities for new forms of environmental governance, grounded in extensive collaboration between people working on related sciences, technologies, urban planning, health, policy, education and the arts. The book describes an array of initiatives (many of them notably experimental and creative) to understand and deal with Delhi's air. The book is also an invitation, calling readers into the collaborative challenges the authors describe."

—**Kim Fortun**, Professor of Anthropology, University of California-Irvine, USA, and author of *Advocacy after Bhopal: Environmentalism, Disaster, New Global Orders* (2001)

"Delhi experiences only about 50 days of clean air on an average every year. This reality (nightmare!) has spawned an entire universe—policy making, awareness generation, technology production, political mobilisation—that seeks to solve this problem. Interesting and important as this might be, there is another world that is perhaps even more interesting and important. Certainly very intriguing! This is the vast but hidden backstage of action, activity and negotiation that simultaneously animates this world of air and its pollution even as it is mobilised constantly. *Atmosphere of Collaboration* is a story of that backstage. Deeply interesting, insightful, provocative and essential reading if the haze has to be lifted!"

—**Pankaj Sekhsaria**, Associate Professor, Indian Institute of Technology-Bombay, India, and author of *Instrumental Lives: An Intimate Biography of an Indian Laboratory* (2019)

Atmosphere of Collaboration

Air Pollution Science, Politics and
Ecopreneurship in Delhi

Rohit Negi and Prerna Srigyan

LONDON AND NEW YORK

First published 2021
by Routledge
2 Park Square, Milton Park, Abingdon, Oxon OX14 4RN

and by Routledge
605 Third Avenue, New York, NY 10158

Routledge is an imprint of the Taylor & Francis Group, an informa business

© 2021 Rohit Negi and Prerna Srigyan

British Library Cataloguing-in-Publication Data
A catalogue record for this book is available from the British Library

Library of Congress Cataloging-in-Publication Data
A catalog record for this book has been requested

ISBN: 978-0-367-44322-1 (hbk)
ISBN: 978-0-367-76531-6 (pbk)
ISBN: 978-1-003-01563-5 (ebk)

Typeset in Times
by Apex CoVantage, LLC

Contents

Figures

Abbreviations

AAP	Aam Aadmi Party
APnA	Air Pollution Knowledge Assessment
AQI	air quality index
BISA	Borlaug Institute for South Asia
CAC	Clean Air Collective
CAQM	Commission for Air Quality Management
CCAPC	Collaborative Clean Air Policy Centre
CEEW	Council on Energy, Environment and Water
CEMS	Continuous Emission Monitoring Systems
CEPT	Centre for Environmental Planning and Technology
CIMMYT	International Maize and Wheat Improvement Centre
CPCB	Central Pollution Control Board
CPR	Centre for Policy Research
CSE	Centre for Science and Environment
CSIR	Council for Scientific and Industrial Research
CSSP	Centre for Studies in Science Policy
EPA	Environmental Protection Agency
EPCA	Environment Pollution (Prevention and Control) Authority
EPIC-India	Energy Policy Institute at the University of Chicago Center in Delhi
EPoD	Evidence for Policy Design
FCRA	Foreign Contribution Regulation Act
HCA	Healthy City Alliance
ICAR	Indian Council of Agricultural Research
ICMR	Indian Council of Medical Research
IG	Indo-Gangetic
IIT	Indian Institute of Technology
JNU	Jawaharlal Nehru University
JOAR	Jharkhandi Organisation Against Radiation

J-PAL	Abdul Latif Jameel Poverty Action Lab
MoEFCC	Ministry of Environment, Forest and Climate Change
MoHFW	Ministry of Health and Family Welfare
NBA	Narmada Bachao Andolan
NCAP	National Clean Air Programme
NCR	National Capital Region
NGT	National Green Tribunal
NISTADS	The National Institute of Science, Technology and Development Studies
NSA	National Science Foundation
OCS	Open Collaborative Science
PHFI	Public Health Foundation of India
PIL	public interest litigation
PM	particulate matter
RWA	Residents Welfare Association
STS	science, technology, and society
TAF	The Asthma Files
UNEP	United Nations Environment Programme
UPE	urban political ecology
VR	Virtual Reality
WAQSA	Women in Air Quality in South Asia
WHO	World Health Organization

Acknowledgements

This book is a collaboration between the authors, which started in late 2015. In this journey, it has benefited immensely from conversations with several wonderful scholars, advocates, and activists in Delhi and beyond. We are grateful to the time many of them gave us, explaining the issue in detail as they saw it from their vantage point.

We thank the following colleagues and students at Ambedkar University Delhi, from whom we have learned a lot over the years: Anil Persaud, Anna Zimmer, Asmita Kabra, Budhaditya Das, Fizala Tayebulla, Praveen Singh, Pritpal Randhawa, Rachna Mehra, Rashmi Singh, Rituparna Sengupta, Shyam Menon, Shruti Ragavan, Surajit Sarkar, Suresh Babu, Syed Shoaib Ali, and Venugopal Maddipati.

This work has been enriched by our association with The Asthma Files (TAF), and, in particular, Aalok Khandekar, Katie Cox, Kim Fortun, Maka Suarez, Mike Fortun, Pankaj Sekhsaria, Priyanka deSouza, and Vinay Baindur.

We thank Lubna Irfan and Rimina Mohapatra at Routledge for their efforts as the book went from an idea to its present form.

Prerna would like to thank her friends Vinisha Singh Basnet, Kritika Maheshwari, Amit Kaushik, Manan Gupta, Saurabh Chowdhury, Aishwarya Kumar, Tim Schütz, Hae Seo Kim, Faye Yuan, Courtney Graves, and Tarek Mostafa Salama for their warmth and strength. She is grateful to Dhritiman Bhuyan for support through 5 years of this effort. Prerna thanks her father Manoj, her mother Seema, and her sister Megha for their faith and love.

Rohit is grateful to the following friends for their support, especially in the exceptional times of the pandemic: Amitabh Kumar, Anu Sabhlok, Kesang Thakur, Persis Taraporevala, Siddharth, Sumantra Chatterjee, Sumitro Chatterjee, and Surajit Chakravarty. He thanks Shruti for companionship and steadfast encouragement.

1 Introduction

Walking towards the venue of an air pollution event, organised by the Centre for Science and Environment with the Delhi Chief Minister in attendance, I run into an expatriate ecopreneur-activist whom I had come to know in the course of research (more in Chapter 5). We enter the packed hall together. Searching in vain for empty seats, we find a perch by a slew of video cameras sent by various news channels. While the CM is a popular figure in general, interest in this event was heightened because the city was four days into the latest episode of vehicle rationing, locally known as "odd/even," and it had already attracted controversy, with the main opposition party having criticised it heavily. As the crowd waited for the CM, I notice my companion exchange greetings with a few individuals who I recognise from other air-related events in the previous months. One of them, another ecopreneur, rises from his seat and walks towards us, gesturing towards his mobile phone. He shows us a graph—on an air quality app—with hourly trends of PM2.5 in Delhi. It had been building up since the morning, and had shot up to about 150 [microgram/cubic meter] by the afternoon. Another acquaintance, who was an advisor to the AAP[1] government as well as a fellow traveller in Delhi's air scene, joins in the conversation. The three of them speculate on the reasons for the spike. "Maybe they've set fire to crops around Delhi," says the expat, an assessment the government advisor doubts. The debate though is quickly adjourned as the CM enters the overflowing hall in Central Delhi.[2]

The interest in the event we open with, and the presence of diverse voices in it, are indicative of the general salience of air pollution in Delhi's public sphere and of its specific trajectory. Environmental advocates work closely with scientists and researchers, and many of them converse directly with the highest levels of the government. Even individuals unaffiliated to well-established organisations have entered the conversation and voice their opinions at public events (both the ecopreneurs in the earlier narrative made interventions during the ensuing discussion that evening). These individuals are, and see themselves as, "armed with data" and an understanding of its implications. They are, in other words, vectors of a splintered expertise that is now available on mobile phones, should anyone be interested. And yet, as is seen in the short exchange described earlier, even when an understanding of the technical variables is shared, there are wide divergences in explanations.

Experiences with critical pollution are, of course, not about Delhi alone. India's economic growth story of the past three decades has been scripted by viewing and drawing on the environment as a limitless cheap resource. Economists, who disproportionately shape public policy in India, pay heed to the environmental Kuznets curve, which argues that countries pass through a phase of environmental degradation as per capita incomes rise, adopting stricter regulations and cleaner technologies only after reaching a certain threshold of economic development. As critical literature notes (Stern, 2004), global discursive shifts towards environmentalism and domestic movements have pushed India towards far more comprehensive laws and institutions for environmental protection than other countries at a similar level of development. There seems a fundamental contradiction at the heart of Indian environmental realities: large-scale degradation coexists with stringent legal regulatory regimes and a decent record of biodiversity conservation.

The Bhopal gas disaster of 1984 remains a watershed in the popular consciousness and environmental governance. The tragedy led to a worldwide awakening around chemical, industrial, and environmental risks faced by already-marginalised communities like Dalits, minorities, and the urban poor, more widely. Stringent laws and regulations, empowerment of state agencies, and right to information towards greater public knowledge of risks were, in the post-Bhopal world, considered essential to effective and just environmental governance (Fortun, 2004). One result of the emergence of this consensus was the widening gulf between environmental conditions in the North and the rest of the world. As the former cleaned up their most severely polluted urban regions, the displacement of hazardous activities elsewhere led to a reconfigured global geography of risk: countries like China, India, and Nigeria emerged as toxic hotspots (Negi, 2020). It became incumbent on these countries to set up their own infrastructures of monitoring and enforcement, a huge challenge that created a significant lag between pollution and its amelioration via regulation and restoration. Meanwhile, corporations continually push back against environmental oversight through well-funded lobbying at various levels of the sociopolitical system (Fortun et al., 2016). This includes a systematic "labour of confusion" via greenwashing campaigns and half-hearted local interventions, exacerbating affected communities' "physical and psychological suffering" (Auyero and Swiston, 2009, p4).

It is in this context that scholars have shown the form of urbanisation taking place in India as environmentally destructive, with issues like land-cover change, air and water pollution, and piling waste that harm public health and disrupt ecology (Rademacher and Sivaramakrishnan, 2013). Dust circulates across urban regions, toxic air and water affect large populations, and landfills constantly leach into neighbouring areas to make many urban regions unliveable. It is another matter that millions have few options other than to carry on living in highly stressed zones since they allow a relatively affordable foothold in the city. A sizable urban population thus subsists while being exposed to multiple toxic matter and gases. These risks extend far into the peri-urban zones, where, additionally, residents must contend with urban detritus like the most hazardous industries while also being sparsely attended to by regulatory authorities (Priya et al., 2017).

Of the environmental risks that characterise the present moment, air pollution is perhaps the deadliest. Toxic air was estimated to account for around 7 million deaths worldwide in 2016 (WHO, 2018). The reliance on thermal energy, promotion of private mobility, explosion of the real estate sector, and the imperative to increase the quantum of agricultural production are together responsible for the highly polluted atmospheres of many urban areas. In the Delhi region,[3] vehicular emissions, construction dust, garbage burning, and stubble fires from the wider region combine with peculiar geography and meteorology to produce a perfect storm such that the city witnesses only about 50 days of clean air—as per national air quality standards—annually (Rampal, 2019). Over 30 million residents of the region are forced to live with respiratory ailments, as the poison builds up inside their lungs, brain and other organs, bringing mortality ever closer. Delhi's place as one of the world's most polluted conurbations (Irfan, 2017) has in turn generated a vibrant public debate on the causes of and interventions to reduce air pollution. Its toxic air is, therefore, a productive vantage point to observe the environmental implications of what has been termed the "urban century," as scientists, policymakers, journalists, activists, and laypeople continually grapple with toxicity.

In a perceptive paper on Delhi's environmental history, Awadhendra Sharan (2006) argues that the "environment is a fluid concept, linking cultures, populations, materials and spaces in specific ways in particular historical conjunctures" (p4910). Sharan describes the shift in state action from an overarching colonial emphasis on nuisance to the attempts to reduce pollution through land-use planning, and from the inclusion of environmental concerns in planning itself, to the judicial assertion of environmental rights. In the 1990s, India's Supreme Court initiated a slew of pollution-abatement actions that transformed Delhi's industrial and transport landscape. After a long-drawn case, discussed in detail in Chapter 4, the court ordered the closure of hazardous industries and the conversion of all public and semi-public automobiles to compressed natural gas (CNG). These interventions led to the loss of livelihood for many workers and auto-rickshaw drivers, many of whom were migrants to the city. Delhi's social justice movements and several scholars strongly critiqued these interventions as elitist, driven by "bourgeois environmentalism" (Baviskar, 2003; Véron, 2006), and complicit with broader attempts to reorder Delhi into a post-industrial "world-class city" to which the working classes and their spaces become superfluous (Ghertner, 2015). Indeed, following the court order, private housing complexes, shopping malls, and parks came up in plots where industries previously stood.

As affected people coped with these shifts and the air visibly cleaned up due to the interventions mentioned earlier (Guttikunda and Gurjar, 2012, p3202), interest in Delhi's air pollution subsided in the 2000s. It was in 2014–15, with a WHO report that placed the city as the world's most polluted capital—a fact denied by the government (Mazoomdaar, 2014)—the US President Barrack Obama's visit to Delhi during the January 2015 smog that was projected to cost him 6 hours of his life (Pearson, 2015), and a New York Times reporter's publicised decision to relocate from Delhi because of pollution (Harris, 2015), that the issue returned to the public eye. As pollution has worsened, the debate too has become more pronounced, especially during the winters when air is at its most toxic (Figure 1.1).

In this book, we track the conversation and interventions related to Delhi's air since 2015. We find that three emergent aspects have assumed importance during this time. First, more than visceral responses and economic calculations,[4] air pollution *science* has assumed primacy in the larger debate, framing critical questions in terms of data accessibility and availability, working with urgency for cleaner air. Second, the toxic air is now widely recognised as a multisectoral and multiscale concern requiring new styles of environmental governance, even as collectives organising for a healthier city seek inclusive means of building momentum. Third, there has been a veritable explosion in the market for commodities like air quality monitors and purifiers to make air pollution a thriving ecopreneurial space.[5] Taking up each of these aspects in the subsequent chapters, we discuss the common threads of interdisciplinarity, experimentation, and collaboration running through the work of the numerous researchers, advocates, activists, and ecopreneurs engaging with air in and around Delhi. In other words, we see, in the contemporary air pollution problem space, an *atmosphere of collaboration* that invites diverse agents with variegated stakes to think and act together towards sustainable and just environmental outcomes.

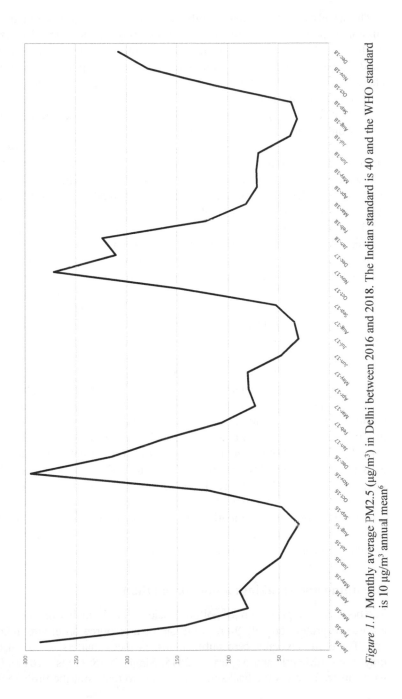

Figure 1.1 Monthly average PM2.5 (µg/m³) in Delhi between 2016 and 2018. The Indian standard is 40 and the WHO standard is 10 µg/m³ annual mean[6]

The widely held belief that credible data are the prerequisite to any discussion on pollution opens the doors to monitoring technologies, dispersion models, and long-term medical studies; to a world of technosciences populated by experts representing a range of interested disciplines from atmospheric chemistry to meteorology and epidemiology to engineering. At the same time, the debate has seen a general diffusion of expertise, and environmental facts today emerge from locations outside state and academic institutions. Nearly each discipline realises the insufficiency of its particular way of knowing the problem, leading to frequent boundary crossings amongst the several air pollution sciences. Their collaborations spread far and wide, bringing in participants from locations around the world, united by the desire to know Delhi's air better to act on it. It takes flexibility, reflexivity, and collaborations to stay relevant.

The institutional architecture through which air is governed is also in constant churning. It involves interactions between the state at many scales (central, provincial, local), multiple regulatory bodies, and different ministries with their own core subject area, but with bearing on air pollution and its consequences. All of this makes governing air a complicated and politically vibrant arena, where larger questions of environmental governance are being debated and contested. Collectives working on environmental issues, for their part, are more attentive to questions of social justice through their depiction of the poor as disproportionate sufferers of air pollution, calling for interventions that do not compromise livelihoods (Narain, 2016). They have, thus, responded to criticisms of elite environmentalism by being mindful of an underlying unevenness in political participation and in the ability to purchase technologies like purifiers to protect oneself from pollution. Even ecopreneurs attempt to reconcile their profit-making from the sale of pollution-ameliorating technologies by working with advocates and activists, and thinking through their actions with regards to the larger public interest. Chapters 3–5 of this book focus on the three communities of practice that have assumed importance in the air pollution problem space: those interested in knowing air scientifically and in the public communication of science; those interested in shaping, implementing, or contesting policy; and those interested in developing techno-solutions.

Breathing and thinking: a note on method

The book is not a primer on air pollution science or air quality management, a guide to protect oneself from pollution, or a resource on environmental law. There are several recent publications on these matters to which one may turn (cf Meattle and Aggarwal, 2018; Singh, 2018; Ghosh, 2019). Our work, instead, enters the backstage, so to say, to tease out the protagonists and stakes, locate them in time and space, and identify the questions, dilemmas, and contradictions that drive the conversation forward. It draws on over 4 years of engaged research, during which time we read several technical reports, vast scientific literature, and—less extensive—social science work closely. We interviewed over 25 scientists, advocates, activists, and journalists and interacted informally with several others. We also observed and participated in tens of panels and other pollution-related events around Delhi.

The book is a product of long-term collaboration and benefits from the skills and perspectives each of the authors brings to the work. One of us (Rohit) is trained in urban planning and geography, and the other (Prerna) has a background in chemistry and in the interdisciplinary fields of human ecology and science, technology, and society (STS), studying sociocultural anthropology currently. The collaboration not only provides the opportunity to think about the various elements of air pollution together but also allows us to engage with aspects of the debate that otherwise would have been foreign to one or the other among us. During this time, we were fortunate to be associated with TAF collective. TAF is a transnational community interested in the links between environmental change, knowledge, and health. We discussed questions, framings, and learnings with the wonderful researchers and practitioners of TAF located across the world through regular online conference calls, and a handful of meetings with the India TAF researchers in Hyderabad, Bangalore, and Delhi. The website of TAF is an installation of Platform for Experimental Collaborative Ethnography (PECE), where researchers can collaborate to share ethnographic material with varied privacy settings and do cultural analysis. Some of our ethnographic material, such as interview consent forms, interview transcripts, and meeting notes, is stored on the website, and can be shared on request.

Conceptually speaking, the work draws heavily from two intersecting but relatively autonomous disciplinary frames. The question at the heart of the book, which is a concern with urban environmental change, is approached productively by the field of UPE. In contrast to some of the hitherto notions of the urban as a one-way process of replacement of nature by the built environment, UPE argues that first, concerns like air, biodiversity, and water are critical to ordinary urban lives, and therefore require attention; and second, if scholars' conception of the urban is expanded from its artefacts to the *process* of urbanisation, then its footprints, routes of production, and by-products emerge as simultaneously social, political, and biophysical questions (Heynen et al., 2006). This conceptual bridging of political ecology and urban studies brings a critical lens to nature in the city, focussing on the larger political economy alongside the many knowledges of and claims around the environment, focusing especially on the interplay of governance and citizenship.

The second frame is the field of STS, which situates practices of scientific knowledge production, technological innovation, and linkages of knowledge with policy and advocacy within a historical and social context. It views the people who engage with science and technology as engaged in cultural and political acts. It invites us to think with such people, understand where they come from and why they do what they do. Bringing UPE and STS into conversation allows us to sketch the entanglements of science, advocacy, policy, politics, and profits as they unfold in Delhi, rather than analysing them in discrete theoretical boxes, and to identify continuities, breaks, and common threads that run through these diverse practices. For instance, the livelihood–environment trade-off remains a powerful frame through which advocacy and interventions are viewed, which makes science— for instance, of source apportionment—a deeply political practice, rather than a cocooned technical exercise. Consider the poster (Figure 1.2)

Figure 1.2 A poster by the Delhi Loha Vyapar Association criticises anti-pollution action citing an unnamed source apportionment study

Source: Photograph by Rohit Negi.

put up around the city by the Delhi Loha Vyapar Association, a collective of people engaged in the iron and steel business, which asks why construction activities around the city get banned during smog episodes when the sector contributes only 1 per cent to the pollution as per an unnamed government report.

Beyond theory, being in the thick of things in Delhi shaped the emphasis and direction of this book. We let the ethnography bring the larger analytics to surface, even when it meant that we had to train ourselves in new perspectives. For us, ethnography means empathetic and critical engagement with differently located agents and their specific worldviews, and as far as possible, to portray the world as it appears to them. Our observations in Delhi distanced us over time from powerful and highly cited critiques from both UPE and STS with which we had begun the research. For example, we never encountered the archetypal scientist who had unquestioning faith in the apolitical and distanced view of science which social studies of science have taken to task. We do not recollect speaking or listening to an environmental activist who did not consider the impacts of pollution or policy on the urban poor. Even the ecopreneurs who sold commodities like purifiers and air quality monitors seemed uncomfortable with the fact that they were benefiting from a public concern, and therefore devoted time and energy to organising for large-scale change. In December 2015, one of us (Rohit) was invited to a brainstorming meeting with a group of environmentalists and open-data activists to initiate a campaign on air pollution. The discussion was lively and a background note prepared by the group (reproduced later) shows just how nuanced some of the air pollution conversation has become in the past decade.

Delhi's residents have been breathing in a toxic cocktail of gases, dust and fine particles lately. Nitrogen Dioxide concentration at most localities in and around Delhi is 4–7 times the levels of 2007–2010 and 2–3 times the nationally prescribed standards. Concentration of fine particles (PM 2.5), which evade the body's filtering mechanisms and enter the respiratory system, are 5–6 times higher than levels usually considered unsafe.

These levels of toxic air have a proven link with increased rates of asthma attacks, breathlessness, persistent cough, and frequent chest infections in the short term, and with incidence of lung cancer with prolonged exposure. Children and populations with pre-existing respiratory illnesses are particularly vulnerable. New scientific evidence suggests that exposure to fine particles is also associated with increased risk of compromised brain functioning, which leads to conditions like Alzheimer's and dementia. Already, many of us suffer on a daily basis, and there is a reported hike in hospital visits due to pollution-related illnesses.

Delhi's air represents a public health catastrophe. It means that concern for the air is a demand for the very survival of all. In fact, vast populations of the urban poor, including construction workers, daily wagers, rickshaw pullers and auto rickshaw drivers are particularly exposed to the toxic air, while also having poor access to quality healthcare. They do not have the option of individualised solutions like indoor air purifiers either. Solutions therefore must be citywide and public in nature. They must also be swift.

We do not deny the scientisation of the larger debate, the middle-class domination in advocacy, or the fact that profits are being generated from the market. Our point is that from scholarly critiques, these issues have become questions-to-think-with for our interlocutors. Critical studies must operate from this recognition rather than offer a distanced critique, no matter how sophisticated its language. As Timothy Morton (2013) says, the "hypocrite understands that she is caught in her own failure" (p148). Since *all* choices are simultaneously inadequate and necessary, the option of remaining untainted is not available to practitioners. The distanced position from where academics often articulate their critique is, in other words, a luxury not available to those who breathe in the toxic city and look to do something about it, no matter how unsuccessful or insignificant these efforts may prove. Our attempt is to consider both the moments of possibility and self-doubt that practitioners face in response to the massive challenge of air pollution, to allow our interlocuters a pause as they go about doing their extremely important work.

Since *we* are writing about them, our presence in this story matters. For most of the last 5 years, while we researched the book, we lived in Delhi. We breathed the toxic air, suffered from some kind of respiratory infection almost every month during the winters each year, popped in antibiotics prescribed by clueless doctors, signed petitions to the government and the courts, and participated in protests around the city. One of us bought an air purifier post-Diwali one year, the other tried out nasal filters. We installed air quality apps on our phones, and checking them multiple times throughout the day became second nature. When acquaintances, friends, and colleagues found out that we were researching these issues, they would seek our advice on which purifier or mask to purchase. Once we had written a few popular articles and interviewed some of the involved protagonists, we started to be recognised and were invited on public panels, interviewed by journalists, and asked to advise an artistic intervention on air pollution. Our status as English-educated and upper-caste academics privileges our engagement with air pollution conversations, and we want to use our access to these spaces to present a reading that accounts for the complexities and varied histories of pollution without being esoteric. We want to make explicit the stakes and commitments of people we talked to and interacted with, including ours, so we can leave space for perspectives beyond our own.

Plan of the book

Chapter 2 tracks the air pollution debate in academic and popular writing in the last decade or so, and outlines the contribution we wish to make to the literature. The chapter parses out what collaborative praxis means to us and how we see it evolve in the air pollution problem space. Finally, it discusses why we chose to focus on three different themes—science and advocacy, governance, and ecopreneurship. The chapter finds connections between these themes.

Thereafter, Chapter 3 reflects on the dilemmas that "expert-advocates" face when they come together to bridge silos and foster interdisciplinary thinking, trying to make sense of their contradictory obligations. As advocacy becomes framed around questions of data accessibility and availability, around life expectancy and productivity, around smaller towns and rural communities, what opportunities and challenges do expert-advocates face, trying to appeal to various publics—the government on the one hand, and citizens on the other? We call this emerging discursive space "science for advocacy," the coordinates for which are being worked out in Delhi as we write.

Chapter 4 discusses the governance of air pollution, looking at the issue both through the administrative architectures and logics, and via the variegated civil society perspectives and actions. Authorities face challenges posed to institutions by the distributed and transboundary nature of air pollution, especially as it encounters rigid administrative structures and ways of seeing and acting on environmental concerns. On the other hand, despite the considerable public awareness and interest in the issue, there is differential participation of citizens in the debate. The chapter considers this unevenness, constitutive of an "atmospheric citizenship," looking closely at the disparate claims and organising strategies for clean air.

As smog hits the region and the media serves wall-to-wall coverage of pollution during the winters, the multimillion and expanding market for anti-pollution commodities, from masks to monitors and purifiers, comes into view. Rather than dismissing this as regressive consumerism, Chapter 5 considers it a product of the specific historical conjuncture where smart urbanism-induced entrepreneurialism meets the rise of urban environmentalism. Tracing the trajectories of individuals who embody these interventions in Delhi shows that these are highly contextual paths at the intersection of personal anxiety, activism, and profit. The entrepreneurial landscape of Delhi's air reveals the deeper processes that lend meaning to subjectivities in activism.

The *postscript* to the book asks what challenges the COVID-19 pandemic poses to the way air pollution is being thought about and responded to by the various protagonists. It sees in the lockdown's clean air a set of possibilities towards a just urban environmentalism, and makes the case for advocates to bring slowness and carefulness in their praxis as the world confronts increased precarity and uncertainty.

Notes

1 AAP or the Aam Aadmi Party is the party in power in Delhi NCT. Its name literally translates as 'party of the common man'.
2 Rohit Negi, fieldnote, 19 April 2016.
3 In this book, unless specified otherwise, we use Delhi to signify not only the administrative unit of the Delhi National Capital Territory (NCT), but the wider region that includes satellite towns like Gurgaon and NOIDA.
4 Thanks to Awadhendra Sharan for this point.
5 Ecopreneurship or ecopreneurialism signifies entrepreneurship in the environmental problem space. More details in Chapter 5.
6 Based on raw data from the US Embassy, New Delhi, compiled by Smart Air. Available at https://smartairfilters.com/en/blog/delhi-pm25-air-pollution-2019-analysis/ (Accessed 20 October 2020).

References

Auyero, J. and Swiston, D. (2009) *Flammable: environmental suffering in an Argentine shantytown*. Oxford: Oxford University Press.
Baviskar, A. (2003) 'Between violence and desire: space, power, and identity in the making of metropolitan Delhi', *International Social Science Journal*, 55(175), p89–98.
Fortun, K. (2004) 'From Bhopal to the informating of environmentalism: risk communication in historical perspective', *Osiris*, 19, p283–296.
Fortun, K. et al. (2016) 'Pushback: critical data designers and pollution politics', *Big Data and Society*, 3(2), p1–14.
Ghertner, D. A. (2015) *Rule by aesthetics: world-class city making in Delhi*. New Delhi: Oxford University Press.
Ghosh, S. (ed.) (2019) *Indian environmental law: key concepts and principles*. New Delhi: Orient BlackSwan.
Guttikunda, S. K. and Gurjar, B. R. (2012) 'Role of meteorology in seasonality of air pollution in megacity Delhi, India', *Environment Monitoring and Assessment*, 184, p3199–3211.
Harris, G. (2015) 'Holding your breath in India', *New York Times*, 29 May. Available at www.nytimes.com/2015/05/31/opinion/sunday/holding-your-breath-in-india.html (Accessed 11 May 2019).

Heynen, N., Kaika, M. and Swyngedouw, E. (eds.) (2006) *In the nature of cities: urban political ecology and the politics of urban metabolism.* New York: Taylor & Francis.

Irfan, U. (2017) 'How Delhi became the most polluted city on earth', *Vox*, 25 November. Available at www.vox.com/energy-and-environment/2017/11/22/16666808/india-air-pollution-new-delhi (Accessed 9 September 2019).

Mazoomdaar, J. (2014) '"Delhi air pollution higher than Beijing" report denied', *BBC.com*, 31 January. Available at www.bbc.com/news/world-asia-india-25957405 (Accessed 27 August 2020).

Meattle, K. and Aggarwal, B. (2018) *How to grow fresh air: India's top experts teach you how to beat air pollution.* New Delhi: Juggernaut.

Morton, T. (2013) *Hyperobjects: philosophy and ecology after the end of the world.* Minneapolis: University of Minnesota Press.

Narain, S. (2016) 'Consequences of inequality for sustainability', *IDS Bulletin*, 47(2A), p113–115.

Negi, R. (2020) 'Urban air', *Comparative Studies of South Asia, Africa and the Middle East*, 40(1), p17–23.

Pearson, N. O. (2015) 'President Obama, the world's worst air just took 6 hours off your life', *Sydney Morning Herald*, 28 January. Available at www.smh.com.au/environment/president-obama-the-worlds-worst-air-just-took-6-hours-off-your-life-20150128-12zl16.html (Accessed 14 April 2020).

Priya, R. et al. (2017) *Local environmentalism in peri-urban spaces in India: emergent ecological democracy?* Brighton: STEPS Centre.

Rademacher, A. and Sivaramakrishnan, K. (2013) 'Introduction: ecologies of urbanism in India', in Rademacher, A. and Sivsaramakrishnan, K. (eds.) *Ecologies of urbanism in India: metropolitan civility and sustainability.* Hong Kong: Hong Kong University Press, p1–42.

Rampal, N. (2019) 'Delhi pollution falls by 4% in 2019 compared to last year', *India Today*, 11 May. Available at www.indiatoday.in/diu/story/delhi-pollution-falls-by-4-in-2019-compared-to-last-year-1616035-2019-11-05 (Accessed 21 July 2020).

Sharan, A. (2006) 'In the city, out of place', *Economic and Political Weekly*, 41(47), p4905–4911.

Singh, S. (2018) *The great smog of India.* New Delhi: Penguin.

Stern, D. I. (2004) 'The rise and fall of the environmental Kuznets curve', *World Development*, 32(8), p1419–1439.

World Health Organization (2018) *9 out of 10 people worldwide breathe polluted air, but more countries are taking action.* Available at www.who.int/news/item/02-05-2018-9-out-of-10-people-worldwide-breathe-polluted-air-but-more-countries-are-taking-action (Accessed 19 August 2020).

Véron, R. (2006) 'Remaking urban environments: the political ecology of air pollution in Delhi', *Environment and Planning A*, 38(11), p2093–2109.

2 Blindspots and collaborations

A map of low-cost air sensors in Delhi was projected on a makeshift wall in the exhibition gallery of the Bikaner House in Delhi. These sensors are located in areas where it is too expensive to install regulatory-grade air monitors. Blindspots of science.

Most bubbles on the map indicating the air quality index were yellow. At least it's not apocalyptic red, I thought. A workstation with a message board directly confronted the makeshift wall. A technician fiddled with virtual reality headsets: "Is it working now?" The visitor adjusted the headsets and nodded. She rotated her head to the right, then to the left. Then, up and down. What was she seeing? I waited for my turn to put on the VR headset. And when my turn came, the headset didn't seem to work. I peered over the technician's shoulder as he made another headset ready for me. I put it on. I was transported near a railway track. A train carrying piles of coal rushed by. Where was I? Korba, the technician replied. I would find out later that I was near the Kusmunda coal mine in the Chattisgarh district of Korba, standing on the land of indigenous communities who suffer intense environmental harm caused by coal mining. I found myself disoriented and took the headset off. I was back in Delhi. I was back to Bikaner House, a former royal residency, now hosting events like this one. But I was not alone near the makeshift wall. A group of excited children gathered around the workstation. One of the exhibition organisers informed me that they were from a Delhi slum. A non-profit had brought them to Bikaner House to make them see that they are not alone in what they experienced in their daily lives. All around us were mural-sized photographs from Mumbai's Mahul, Chhattisgarh's Korba, and Chennai's Ennore. The photographs accompanied stories of people without a lot of resources fighting for justice and recognition. Blindspots of advocacy and governance.

What messages would the children like to leave for Delhi's Chief Minister and India's Prime Minister, asked the activist showing the children around. They scribbled on the message board with ink made from Delhi's particulate matter. These special pens were made by entrepreneurs with a social message in mind: pollution can do political work. The pens filling blindspots of industry. In Bikaner House.[1]

We begin the chapter with this fieldnote to acknowledge the work of recognising and filling blindspots that environmental advocates do. The recognition of blindspots makes collaboration the primary mode of praxis around air pollution. It is insufficient to let data speak by itself, or wait for reports on pollution to find their way into policy. Stories of harm and complexities of pollution are increasingly close at hand. Delhi too cannot stand alone. It requires Mahul and Ennore to tell the larger story about environmental destruction. The organisers of the *Breathless: Documenting India's Air Emergency* exhibition understand these challenges deeply. Environmental journalist Aruna Chandrasekhar collaborated with photographer Ishan Tankha to produce the mural-sized images with captions referencing social justice. Documentary filmmaker Faiza Khan directed the VR film "The Cost of Coal," which visitors could experience through the headsets. Respirer Living Sciences, founded by Ronak Sutaria, whom we will encounter in the subsequent chapters, produced the low-cost sensors.

Delhi-based Help Delhi Breathe and the pan-India Clean Air Collective (CAC) supported the exhibition. Two discussions inaugurated the event, one between representatives from three main political parties of Delhi, and another between a waste worker, a lung cancer survivor, and his family. Composer Ankur Tewari performed a new song on his guitar about not being able to breathe in Delhi's smog. In a unique ceremony, the political representatives gifted the waste worker, the cancer survivor, and his family framed pieces from a 3-D artificial lung that members of the Delhi-based Lung Care Foundation had installed outside Delhi's Gangaram Hospital the previous year. The lung had turned black in 48 hours and made national news.

What does it mean for environmental advocates to come together and fill blindspots of knowledge and practice? The people we begin our fieldnote with have identified many blindspots: how conversations on air revolve around Delhi or other metropolitan cities, how enough stories are not told of people who are forced to live and work in conditions that give them disease and produce far-reaching toxic legacies. They know that Delhi alone cannot reduce its air pollution. It needs coordination with state governments of Haryana, Punjab, and Uttar Pradesh. They struggle, while using N99 masks, for a time when masks may no longer be necessary.

When we started our work on Delhi's air in 2015, most of the partnerships and collaborations we write about in this book had just begun. Some did not exist at all. Environmental advocates with previous sustained engagements were rethinking their positions. Politicians brought up air pollution in parliamentary debates. More than ever, scientists began to ask critical questions about knowledge production and dissemination. They wondered if their work was being heard or understood by people and governments.

Industries and businesses looked to partner with non-profit organisations to develop emission standards. In all of this, questions of accountability and justice became vital. Yet, as environmental advocates assembled in Bikaner House's gallery to discuss social and environmental justice, they occupied a building with visible colonial attachments. Cultural events of public interest are often hosted at the House, but the building's enclosed, grand facade invites few within. Even though the exhibition did critical comparative analysis, it did that in Delhi. Even though the children wrote messages mostly in Hindi, the photographs' captions around them were written in English. These exclusions escape no one. So how do environmental advocates—doing science, activism, governance, or entrepreneurship or all of these—grapple with these blindspots with their delimited fields of view?

We write the subsequent chapters with this question in mind. As we mention in the book's Introduction, whether in air pollution science, activism, governance, and entrepreneurship, we are committed to empirically studying what is going on as an example of doing situated conceptual work. Often, we make decisions about what kind of positions the people we study and work with occupy before research actually starts. Sometimes, we have a specific community in mind when we plan our research. In our case, we found it difficult to define a community and often wrestled with how we would label our interlocutors while writing about them. Would we stick with their specialised expertise (scientist, activist, entrepreneur)? Or would we choose to call all of this advocacy?

We respond to this by choosing the broad descriptor "environmental advocate" if we refer to them together. Doing advocacy means articulating one's position on something to someone else in hope of change. It may also mean refusing articulation. As researchers, we listened to what people were saying as well as to their silences to grasp blindspots in thinking about air. But we also wanted to understand their specific trajectories and commitments. Where appropriate, we have tried to evolve descriptors to articulate these trajectories and commitments. For example, in Chapter 3, when we talk about people traversing boundaries of air pollution science and advocacy, we use the term "expert-advocates." In Chapter 5, when we talk about people who have made air their business, we use the term "airpreneur." In both these cases, we show why we use these terms and how they change the analysis of Delhi's air. Therefore, as much as we focus on how people grapple with various blindspots of air pollution in Delhi, we hope that readers get a sense that there is no one type of environmental advocate when it comes to Delhi's air. We grasp these blindspots together with our interlocutors and hope that our limited intervention makes sense to them.

The blindspots of air's present demand many voices and contributions. In this chapter, we write on what has been said about air so far, particularly in the Global South. Learning from early perception studies of communities made to suffer environmental harm to the popular non-fiction books that give a comprehensive transnational overview of the science and politics of air, we ask: How are the social sciences called upon to respond to the blindspots of air's present? What is expected of us and how do we respond? These questions take us towards collaboration, our second concern in this chapter. We end by writing about what we can learn from studies of collaborations for environmental research and advocacy. As the themes of experimentation and collaborative practice remain persistent in our research, we hope that these insights can ground us as we go forward.

Towards the present: perception studies to critical research

An early intervention from the social sciences into environmental research was the field of risk perception studies. It emerged when public opinion surveys in the 1950s and 1960s found that communities stressed by toxics articulated a sense of invulnerability (Bickerstaff, 2004). To experts, this lack of concern was ignorance and irrationality, a "perceptual bias" explained by the lack of scientific awareness, to be corrected through strategic transfer of bias-free scientific knowledge. This assumption is at the base of the imagined ability of technoscientific knowledge as essential to transforming behaviour. We heard often in conferences and seminars that people of Delhi did not care about wearing a mask out of a sense of possessing invulnerable lungs and bodies. What do the perception studies tell?

Saksena (2011) reviews the air pollution perception literature with a focus on the type of studies that should be conducted in the Global South. He argues that interdisciplinary teams of physical and social scientists should conduct longitudinal studies, especially on how perception is related to poverty and livelihoods. The global health organisation Vital Strategies took up this challenge and authored a report (2019) that analysed around 500,000 pieces of social media content across ten countries in eight languages. It even ranked public influencers for each year. In 2018, Delhi Chief Minister Arvind Kejriwal emerged as the top social media influencer out of the ten countries. Not a single public health official or advocate was an influencer. They found that people increasingly talked less about masks and more about long-term solutions such as energy transitions, transportation changes, and waste management. Vehicular emissions were the number one source discussed in social media. The report interpreted this as people not focusing enough on important sources such as household fuels, waste burning, and power plants. Posts framed around children's health and climate change evoked the most response, which the report interpreted as "emotionally appealing content."

Closer home, a perception study was conducted in 2019 by the Delhi-based organisation United Residents Joint Action (URJA), an advocacy group of residents' welfare associations (RWAs) across Delhi (URJA and ARK Foundation, 2019). A total of 509 people from ten locations across Delhi were interviewed. Among other findings, the report says that 71 per cent of the respondents were not satisfied with Delhi's air quality, but 89 per cent did not know that there was an air pollution monitor in their locality. Most had not seen the LED screens with air quality information at busy traffic intersections, and of those who had, many did not understand what the colours and numbers meant. Both surveys are not examples of risk perception research the way Saksena (2011) thinks they could be done. But as efforts trying to make sense of lay knowledges, they remind us that people are acutely aware of the toxicities they face every day, even if their ways of knowing may not correspond to technoscientific vocabularies.

URJA's co-founder Ashutosh Dikshit was one of our interlocutors. In our conversation, he reflected on this dilemma:

> Because of media coverage, everyone knows that the air is polluted. That it has PM2.5. But have we seen PM2.5? People talk about air in RWA meetings, but they have a different relationship with it. They'll see garbage and plastic burning, emitting smoke. Dust from construction is entering people's homes and causing asthma. Ask Dr. Guleria of AIIMS about how many patients are worried sick about dust. The situation is pathetic. In Mundka, people see that sewer water overflows, enters their homes and is seeped into food. People see and understand.
> (Personal communication, 2017)

Ashutosh Dikshit implies here that people do not assimilate communicated air quality information passively but infuse it with their local and experiential knowledge, also observed by scholars elsewhere (Bush et al., 2001; Bickerstaff, 2004; Bickerstaff and Simmons, 2009; Cupples, 2009). This may seem obvious to many but people's local experiences are stubbornly absent in many technoscience-oriented air pollution conversations. To the authors of the influential IIT Kanpur source apportionment report on Delhi's air (2016), for example, individuals are invisible, not even showing up in the section about Delhi's demography, which only lists the city's physical geography, population, size, key industries, literacy, and number of languages—that is, standard general knowledge information. Farmers make an entry, but as people who indulge in wasteful practices that must be banned outright. Supposedly a study in air quality management, it relies on optimum modelling scenarios for pollution control options, suggesting that if burning of municipal solid waste, biomass, and coal is banned, pollution loads from those sources would entirely vanish. We write more about expectations, demands, and dilemmas of doing air pollution science and public health in Chapter 3, but these exclusions are why statements like Dikshit's remain important for environmental advocacy. Dikshit was not a scientist and he lived in the elite neighbourhood of Delhi's Defence Colony, but he had a fundamentally inclusive notion of doing air quality and public health communications work (more in Chapter 4).

Besides risk perception studies and public opinion surveys, our interlocutors also paid attention to non-fiction publications like Beth Gardiner's *Choked* (2019), Tim Smedley's *Clearing the Air* (2019), energy policy researcher Siddharth Singh's *The Great Smog of India* (2018), and air pollution scientist Gary Fuller's *The Invisible Killer* (2019). These publications have much in common apart from being published around the same time. All focus on the silence and salience of air pollution as an invisible enemy that must be defeated. Fuller grieves that there is no memorial for people who died in the London Smog of 1952. Gardiner stresses on the invisibility of air pollution materially and politically, noting its immense health and economic costs. Singh asserts that air pollution may have killed more people in India than terrorism but has received far less attention. Smedley offers a multitude of data to portray air pollution as a global emergency.

They, however, differ in their geographical scope and the kind of strategies they outline. Singh's book connects air pollution to the histories of India's industrial, agricultural, energy, and transport sectors, and ends with ten key strategies, aspiring to provide a comprehensive account of India's air pollution with an eye to energy futures. Fuller's book focuses mostly on Europe and the USA. Gardiner, on the other hand, goes on a global tour of air pollution, writing stories from London, California, Berlin, Malawi, Poland, Delhi, and Beijing. The first three places become examples of how clean air policies can be evolved through a combination of science and advocacy. These books are useful because they help advocates do necessary comparative analysis together with the perception surveys. They allow a shared transnational vocabulary around which further environmental advocacy might take place. They serve a moral purpose in unsettling readers and making them recognise their apathy and complicity in making air pollution a global concern. Siddharth Singh, for example, is convinced that redistribution of public infrastructure is necessary:

> Within income inequality there have been a lot of movements in the past century, past several decades where you know, redistribution

through welfare means and so on. Now, you cannot distribute clean air, right? The only way you can address this issue is by ensuring that the poor have access to the same kind of infrastructure that can lead to cleaner environments.

(Personal communication, 2019)

These concerns have been of interest to UPE. It argues for reimagining cities as socio-natural networks of entities, materials, and ecologies generated by "urban metabolism," that is, the constitutive inflows of resources and outflows of waste. Controlled by a minority, these processes tend to marginalise subaltern communities on whom environmental externalities are often imposed (Swyngedouw, 1996). This marginalisation is accentuated by the intersections of class, caste, and gender, thus reinforcing or reshaping existing hierarchies (Doshi, 2016). Despite this tradition of critical research, a lacuna in conceptualising air as being integral to urban natures has been observed by UPE scholars (cf Graham, 2015). This might be, in part, because it is difficult to fix air's complex physico-chemical properties to sociopolitical structures. Moreover, it is only during peak-pollution events that air is rendered visible and enters the public discourse by virtue of its viscerality. At most times, "capturing" air requires complex monitoring devices and specialised knowledge, which effectively confine discussions of air and its pollution to disciplinary spaces of the environmental sciences and public health (Beaumont et al., 1999). Though such studies suggest differentiated vulnerabilities to pollution, they do not adequately problematise the conditions of their production or comment on their sociopolitical implications, choosing to focus on ways to facilitate better communication of air quality in the hope of combining top-down scientific expertise with urban planning and governance (see Chapter 3). It is in this backdrop that calls to consider expertise, policy, and activism related to air as socially produced and power-mediated have emerged in UPE (and STS) literature (Choy, 2011).

While we appreciate the popular publications on air cited earlier, because they fill a gap in global air quality and science communication, as the UPE critique signals, we find that more attention is needed to specific histories and stakes of environmental advocacy rather than framing it as an open space on which the reader may etch a new story of action. Even as the analysis of toxic air gets ever more refined, very few publications delve into the context-specific trajectory advocacy, or how debates around who should do what unfold across different locations. This book is a story of what asking and thinking through this question looks like in Delhi in the late 2010s. That is why we chose to begin with the kind of collaborative practice that the exhibition attempted. It not only talked in a shared transnational vocabulary by terming air pollution an emergency but also engaged particular stories of environmental advocacy.

Practicing collaboration

During our research, we faced at least four kinds of expectations from our interlocutors. The first was about the type of data we were collecting and how we would use it. What was our sample size? Was it enough? If we archive our conversations and reflections in a collaborative digital platform, who would have access? The consent forms people signed when they generously gave us their time often prompted these questions. The second expectation was knowing enough about air pollution science, activism, governance, and entrepreneurship: What did we know about the chemistry of air pollution or how models worked? Why did we think a particular air monitor was placed in one neighbourhood and not another? The third expectation was communicating what we learned from other people. What did we think of a particular non-governmental organisation? Did we think that a politician's speech made sense? The fourth expectation was communicating science and advocacy efforts. In a conference on air pollution monitoring at a technical university one of us went to, a scientist called to the audience: Where are the social scientists? Would they please help in communicating our work to the public?

Just as we discuss the blindspots of Delhi's air, our interlocutors gauged *our* blindspots and pushed us to think about our own research and practice. We were at once researchers, facilitators, translators, communicators, and advocates. Their expectations about data accessibility and curation, about facilitating and translating insights from one domain of practice to another, about communicating work told us to worry less about the entry of social sciences into conversations about air. The popularity of perception studies and non-fiction publications told us that there was enough interest in conversations beyond technical fixes. People wanted to know about the history and politics of environmental advocacy, about the dilemmas advocates faced, and what to do about them. Conversations on air in Delhi cross boundaries as frequently as air itself. Our views were, for instance, listened to by scientists even as we were invited to work with an artists' collective on a long-term arts-based engagement with air pollution. Our challenge was to think through the specific arguments and positions we were to foreground as ethnography opened up possibilities of doing advocacy and communication. With these expectations placed on us as social science researchers, we now briefly write about how we approach collaboration, the underlying theme of this book. We explore the nuances of each domain expansively, for which this chapter serves as a grounding text. We are inspired by our interlocutors who continue to do collaborative work in face of limited resources and global inequalities of authoritative science and advocacy. They do this work despite the class, gender, and caste exclusionary practices that abound in science, activism, governance, and entrepreneurship.

Collaborations are not new for environmental research and activism, far less for research and activism in general (Derickson and Routledge, 2015). The argument that very little is achieved by people, ideas, and things placed in silos is convincing. Many of the collaborations we describe are among researchers who are not located in the same geographic region or the same sphere of work. In collaborating across divides, they document environmental knowledge and environmental injustice, whether they speak about their work in this way or not. They wish to transform the practice of research and activism by contributing new words and tools. They might have differing backstories. They might not share political beliefs and have disparate ways of thinking. Their intention might be straightforward, such as reducing air pollution, but they mostly acknowledge that the path to getting there is complicated. While we do find some alliances that conform to Margerum's (2008) typology, that is, action (community-based), organisational (interest groups), or policy (regional policymakers) collaborations, in Delhi's air pollution space, we find cross-cutting alliances, which often bring together community-based groups, research organisations, and even government scientists. Other collaborations cut across different epistemic communities such as environmental scientists, atmospheric scientists, public health experts, and medical practitioners. These are the kinds of collaborations we encountered, and we want to learn from their practices in this section.

Collaborations within the sciences usually end with co-authorship of the research output and evolving a mechanism for sharing disparate databases or scientific tools. They might extend to educating and training scholars, and sharing funding opportunities for future research. Partners located in Southern institutions are supposed to open up their databases and archives to supposedly global institutions which are, in fact, Northern. In turn, the latter offer co-authorship and funding opportunities. Okune (2019) notes that even when partners located in Southern institutions publish with their Northern peers, they do this in journals with paywalls. If they want their work to be more accessible, the additional labour they do is not rewarded by academia (Negi and Ranjan, 2020). It is not surprising to find peer-reviewed articles about environmental injustice in India without a single reference from a Southern institution. This is common in the social sciences too. Such omissions are not accidental but a result of failing to do due diligence. Collaboration certainly doesn't mean only citation but giving academic credit is crucial because academia remains precarious. How should collaborations attribute intellectual labour?

Some collaborative practices have a more inclusive notion of challenging the status quo, such as the Open Collaborative Sciences (OCS) movement (Okune et al., 2016). OCS aims to disrupt the extractive status of the North–South institutional collaborations by building shared research commons as a resource for researchers and communities. It will consider many types of knowledge as equally valuable while recognising the history of injustice done to knowledge produced by Southern institutions and developing mechanisms to establish intellectual and researcher equity and accountability. We observe other movements towards this direction. There is, for example, an increasing focus on institutionalisation of community–researcher partnerships, such as California's Assembly Bill 617, which mandates air monitoring systems in priority areas chosen by evaluating personal and community vulnerability to air pollution in an airshed (Wong et al., 2018). The terms and necessities of these systems have to be developed by community members and researchers together. They will be community-owned and -operated hardware and software with plans for limited financial burden on the community. Public officials in California are already representing this as a global model for public health action (English et al., 2017).

Our interlocutors are watching these collaborations in California closely, as low-cost community air monitoring is now the material infrastructure on which collaborations get imagined. Many of our interlocutors are involved in low-cost air monitoring projects in Delhi or elsewhere. Others are finding ways to include such projects in their practice. But this does not mean that they think community air monitoring will be a silver bullet for public health. Environmental science researcher R Subramanian cautions:

> It will be interesting to observe what happens to the data, how they are translated, what action happens, who is held accountable? People who are industry-friendly might look for holes in data. Can the California EPA push for action when that happens? What happens after funding? Where are the data stored? For many people, innovating means the internet of things. As soon as you get a dashboard open, the work of action is done.
>
> (Personal communication, 2019)

Besides questioning the virtues and commitments of community air monitoring networks, air pollution science researchers are talking more concretely about challenging the status quo. The podcast Atmospheric Tales is a collaboration between Shahzad Gani, Pallavi Pant, and Priyanka deSouza— three early-career researchers of the Indian diaspora in the air pollution sciences who are committed to environmental justice and to pushing back against hierarchical power structures in science. Priyanka also co-hosts the Decolonizing Science podcast at MIT Colab Studio, where one of us was a panellist on a roundtable to talk with people whose work we've followed for some time now, discussing how greater interdisciplinary attention to the blindspots of the sciences, its links with policy and politics, and appropriate framings of justice may move towards more inclusive practice.

These conversations were also part of the annual meeting of the International Society for Exposure Science (ISES) in September 2020. One of us participated in the panel "Decolonising Air Pollution Science: Understanding the Power Structure." The panel was organised by OpenAQ, an open data collaboration which integrates data with questions of social justice, using the term "air inequality" to do their advocacy work. We were asked about what needs to be done to decolonise air pollution science. An intentional conversation to challenge the status quo of science made it possible for a social science researcher to communicate their work and commitment to a largely scientific audience. Our response was first about the importance of acknowledging the blindspots of environmental advocacy and governance. Second, it was about recognising that decolonisation is incomplete without redistribution of land, resources, and public infrastructure. We should question if community–researcher collaborations can be ethical without thinking about who owns not only a community-owned air monitoring system but the land on which that system gets built.

A particularly hard question for collaborations is how to build and sustain trust. Usually, trust is established before collaboration. But in the highly dynamic conversations around air, people have multiple allegiances and obligations which might conflict with each other, as we show in the next chapters. Kali Rubaii (2020, p213) offers a way to think differently about trust:

> What does trust look like when the future is neither a cumulative project nor a leap made in good faith? What does it mean to act on trust without confidence, a trust that neither contributes to, nor relies on, solidarity?

She studies how people who smuggle medicine in ISIS-controlled Iraq to the city of Mosul build trust networks with both allies and adversaries. They articulate a moral pragmatism about other people, even adversaries, choosing life every time an opportunity to trust is offered to them.

For Delhi's air and environmental issues in general, this comparison might seem far-fetched. But the comparison lays out a practical approach for collaborations. It says that while building trust is crucial, enough people have faith that enough people will choose to do the right thing. This seems counterintuitive to calls for building solidarity first and collaboration later, but allows working together for people who have very different ways of thinking and doing things. It allows an entrepreneur who benefits from toxic air by selling masks or air purifiers to work with an environmental justice activist. Their trajectories and intentions might be different, but temporarily, they will come together and speak for clean air.[2] Such uncomfortable collaborations characterise Delhi's air today. Our task in this book is to tease such characterisations and their dilemmas out.

We conclude this chapter by inviting readers to think about these questions as they read what follows: What access to advocacy and educational resources do communities have that practitioners can form synergies with? What is the communities' relation to policy, the judiciary, and public infrastructure? What mechanisms of trust and accountability can environmental advocates evolve? These questions are not ours alone but have evolved in the course of conversations with our interlocutors, paying close attention to what they were saying and not saying. In the subsequent chapters, we detail not only the specific historical moment at which these possibilities open up but also the challenges advocates face towards participatory and collaborative practice.

Notes

1 Prerna Srigyan, fieldnote, 5 June 2019. We have archived this exhibition as an example of air pollution advocacy on The Asthma Files platform.
2 Rubaii calls this "enunciatory trust," building from Kim Fortun's (2001) use of enunciatory communities, where people with different stakes converge temporarily without completely agreeing on their goals and stakes.

References

Beaumont, R., Hamilton, R. S., Machin, N., Perks, J. and Williams, I. D. (1999) 'Social awareness of air quality information', *The Science of the Total Environment*, 235(1–3), p319–329.
Bickerstaff, K. (2004) 'Risk perception research: socio-cultural perspectives on the public experience of air pollution', *Environment International*, 30(6), p827–840.
Bickerstaff, K. and Simmons, P. (2009) 'Absencing/presencing risk: rethinking proximity and the experience of living with major technological hazards', *Geoforum*, 40(5), p864–872.
Bush, J., Moffatt, S. and Dunn, C. (2001) ' "Even the birds round here cough": stigma, air pollution and health in Teesside', *Health & Place*, 7(1), p47–56.
Choy, T. (2011) *Ecologies of comparison: an ethnography of endangerment in Hong Kong*. Durham, NC: Duke University Press.

Cupples, J. (2009) 'Culture, nature and particulate matter: hybrid reframings in air pollution scholarship', *Atmospheric Environment*, 43(1), p207–217.

Derickson, K. D. and Routledge, P. (2015) 'Resourcing scholar-activism: collaboration, transformation, and the production of knowledge', *The Professional Geographer*, 67(1), p1–7.

Doshi, S. (2016) 'Embodied urban political ecology: five propositions', *Area*, 49(1), p125–128.

English, P. B. et al. (2017) 'The Imperial County community air monitoring network: a model for community-based environmental monitoring for public health action', *Environmental Health Perspectives*, 125(7), p074501.

Fortun, K. (2001) *Advocacy after Bhopal: environmentalism, disaster, new global orders*. Chicago: University of Chicago Press.

Fuller, G. (2019) *The invisible killer: the rising global threat of air pollution-and how we can fight back*. London: Melville House.

Gardiner, B. (2019) *Choked: life and breath in the age of air pollution*. Chicago: University of Chicago Press.

Graham, S. (2015) 'Life support: the political ecology of urban air', *City*, 19(2–3), p192–215.

IIT Kanpur (2016) *Comprehensive study on air pollution and Green House Gases (GHGs) in Delhi*. Kanpur: Indian Institute of Technology.

Margerum, R. D. (2008) 'A typology of collaboration efforts in environmental management', *Environmental Management*, 41, p487–500.

Negi, R. and Ranjan, A. (2020) 'Delhi's fight against air pollution could get a boost if science was decolonised', *Scroll.in*, 11 November. Available at https://scroll.in/article/976361/delhis-fight-against-air-pollution-could-get-a-boost-if-science-was-decolonised (Accessed 11 November 2020).

Okune, A. (2019) 'Decolonizing scholarly data and publishing infrastructures', *LSE Impact of Social Sciences Blog*. Available at https://blogs.lse.ac.uk/africaatlse/2019/05/29/decolonizing-scholarly-data-and-publishing-infrastructures/ (Accessed 2 April 2020).

Okune, A. et al. (2016) 'Tackling inequities in global scientific power structures', *The African Technopolitan*, 4(1), p128–131.

Rubaii, K. (2020) 'Trust without confidence: moving medicine with dirty hands', *Cultural Anthropology*, 35(2), p211–217.

Saksena, S. (2011) 'Public perceptions of urban air pollution risks', *Risk, Hazards and Crisis in Public Policy*, 2(1), p1–19.

Singh, S. (2018) *The great smog of India*. New Delhi: Penguin.

Smedley, T. (2019) *Clearing the air: the beginning and the end of air pollution*. London: Bloomsbury Sigma.

Swyngedouw, E. (1996) 'The city as a hybrid: on nature, society and cyborg urbanization', *Capitalism Nature Socialism*, 7(2), p65–80.

URJA and ARK Foundation (2019) *A study on Delhi's perception and accountability on air pollution*. Available at www.indiaspend.com/wp-content/uploads/2019/01/RTI-Survey-Report-on-Delhis-GRAP-URJA.pdf (Accessed 9 July 2020).

Vital Strategies (2019) *Hazy perceptions: public understanding of air quality and its health impact in South and Southeast Asia, 2015–2018.* Available at www. vitalstrategies.org/wp-content/uploads/import/2019/03/Hazy_Perceptions.pdf (Accessed 13 August 2020).

Wong, M. et al. (2018) 'Combining community engagement and scientific approaches in next-generation monitor siting: the case of the Imperial County community air network', *International Journal of Environmental Research and Public Health*, 15(3), p523.

3 Science for advocacy
Thinking with expert-advocates

From December 2017 to August 2018, the Centre for Policy Research (CPR) in Delhi convened a ten-episode public seminar series, *Clearing the Air: Air Quality Regulation and Governance in India.* The organisers of this seminar series are climate change, environmental, and energy policy researchers informed by challenges of coordinational capacity and committed to public outreach. They look to shift the contours of conversation around Delhi's air from seasonal and episodic peaks to a more persistent concern throughout the year. In this series, CPR invited experts from science, advocacy and governance to speak on how they interpret the problem of air pollution for Delhi and broadly, for India. Scientists, economists, lawyers, planners, journalists, consultants, and activists interacted over a year about how to understand and intervene in this stubborn, perplexing problem.[1]

On 23 February 2018, an agronomist, an agricultural geneticist, and a veteran journalist were part of a conversation with a farmer from the district of Panipat in Haryana, a neighbouring state of Delhi with which it shares borders and severe air pollution. The panel was about crop residue burning, known to intensify air pollution during winter. Many farmers in Delhi's bordering states of Punjab, Haryana, and Uttar Pradesh burn paddy residue to clear the ground for wheat cultivation at the start of the new sowing season. The smoke crosses state boundaries and causes an uproar in national and local media. Climate change, agricultural, and energy researchers around the world are trying to figure out how to intervene. One of them is Dr. M.L. Jat, a senior cropping systems agronomist, who talked about the Happy Seeder, a tractor-mounted machine that uproots paddy residue, sows wheat, and deposits the residue back into the ground as mulch. The state government of Punjab gives subsidy on the equipment of up to 80 per cent to farm cooperatives. Dr. Jat was convinced that farmers should abandon old farming practices and embrace new ones. "What does a farmer want? A farmer wants options," he said.

In Jat's story, the Happy Seeder was offered to the farmers as an indigenously developed, sustainable technological solution to improve both rural and urban air. The farmer in conversation, Pritam Singh Hanjra, questioned the options. His story began with migration from Pakistan to Indian Punjab, and he reflected on the Green Revolution's legacy of mimicking water-intensive paddy cultivation practised in the Indo-Gangetic (IG) belt that caused intense waterlogging and salinity in Punjab's farm fields. He talked about reading textbooks published by agricultural universities of Punjab and Haryana and experimenting to reduce water needs of paddy farming. Though not trained through conventional modes of acquiring technoscientific knowledge, Hanjra presented himself as an "organic scholar" in the Gramscian sense, grounded in his field experiments, driven by necessity to innovate. Jat frequently interrupted Hanjra's story to bring out caveats in those field experiments. "I am trying to provoke him," he explained to the audience.

But not much provocation was needed for Hanjra to articulate what he thought was the nature of the problem. "Air pollution is not only a farmer's problem; it is humanity's problem . . . Happy Seeder is not the only solution. What farmers need is a task force that prioritises the problems of farmers across India," Hanjra reasoned. He thus tied Delhi's air to the nationwide agricultural crisis. Hanjra deftly fielded Jat's questions, welcoming collaboration and learning with agricultural scientists in the future. Jat's frequent interruptions, though, caused uneasiness in CPR's seminar room, as the audience of mostly researchers and activists realised that even though equal space had been granted to the scientist and the farmer, it had done little to flatten the hierarchy between them. At one point, someone from the audience asked Dr. Jat, "Why don't you let him speak first?"

This was not the first time we would encounter Hanjra in Delhi. When Prerna went to observe another dialogue between scientists, industry representatives, activists, and consultants, held in the conference room of the World Wildlife Fund's Delhi office, Hanjra was in the audience. The event was organised by the Council on Energy, Environment and Water (CEEW), a non-profit research institution. The moderator, an energy policy researcher, asked questions from the farmers in Hindi and translated their answers to the rest of the audience in English. There were a couple of participants who would most probably not have understood Hindi, one from the Swiss Development Agency and another from the International Institute for Applied Systems Analysis of Austria. CEEW, like CPR, advises the Indian Government on policy and receives funding from many sources—governments, non-profits, foundations, trusts, corporate philanthropy (including Exxon-Mobil and Shell), and multilateral development agencies. As the moderator continued to translate, one researcher in the audience raised his hand and asked, "Why don't we have the entire discussion in Hindi? After all, who are we working for?"

In the previous chapter, we discussed how scholars writing on environmental politics in the Global South have noted the difficulties for environmental experts and advocates to include the people they speak for in decision-making and solution-finding efforts. For Delhi's air, specifically, scholars like Baviskar (2003) and Véron (2006) have characterised prevailing advocacy as disconnected and elitist (see Chapter 4). They contend that several interventions to clean Delhi's air have further segregated public space and life in the city. Communities pushed to the city's periphery are forced to live near landfills, power plants, waste treatment plants, industrial corridors, and transport hubs, reinforcing environmental risks.

The organisers who invited Hanjra to those meetings realise this exclusion. They are experienced advocates for environmental justice connected to national and transnational networks. They offer legal advice and consultancy to environmental movements and activists. They walk around with air monitors in their hands in toxically stressed neighbourhoods. They interact with policymakers of contrasting political affiliations. Some, like Dr. Jat, come from modest backgrounds and seek to make themselves useful to public causes. In the absence of a full-fledged dismantling of power and hierarchy, increasingly, they are the only people in the room with a thought to question the status quo.

This chapter builds on the facts of exclusion but shifts analytic attention to think *with* the contradictions experts and advocates confront in their attempt to be seen, heard, and taken seriously—both by the state and the public. Our interlocutors are anxious to be kept in the loop and to include the public they speak for in the loop. We want to understand what aids hearing and learning. As scholars within the social sciences and humanities, we have stakes here too. If our status as English-educated and upper-caste academics privileges our engagement with air pollution advocacy, we want to use our access to these spaces to bring out the dilemmas of collaboration for advocacy. Our intervention in these spaces is to make explicit their and our stakes and commitments. We ask: What contradictions and dilemmas do our interlocutors face in their attempt to be kept in the loop and to keep the public in the loop? What styles of expertise and advocacy do they draw on? What gaps and caveats should they pay attention to?

In this chapter, we approach these questions by first locating our work theoretically within STS and the anthropology of science and technology. In the last 50 years, these fields have established that questions of science and technology are crucial to society, as much as questions of politics and power are crucial to science and technology. These disciplines teach us to grant experts and advocates intricacies of their thinking, and think with them to interrupt their and our own patterns of thinking and doing. We start by rec-ognising that scientists and advocates are asking difficult questions about our present which might not bear familiar, certain, or comfortable answers. Our main argument is that when advocates assemble to bridge silos, they co-produce new spaces and communities to move beyond their otherwise varied stakes and obligations.[2] We show how existing traditions of science and advocacy have worked towards effective regulation on the one hand, and greater public understanding on the other, but increasingly coalesce towards what we call "science for advocacy," with its particular possibilities but also caveats. The first caveat is that of time. Science for regulatory pur-poses and public understanding require particular engagements with time. Our interlocutors, whom we call "expert-advocates," are forced to work under compressed time. While newer institutions take advantage of these changing conditions and compressed time, the enforced urgency forecloses other institutional engagements, such as long-term public health research. The second caveat is that as our interlocutors shift attention beyond Delhi, questions and critiques of political economy enter into their articulations, even as they remain largely absent from discussions of and in Delhi. In the following section, we contextualise science/advocacy linkages of the pre-sent before describing their more recent reconfigurations.

Styles of expertise and advocacy

Placing advocacy

Scholars interested in science and technology from and in the Global South often mediate between multiple literatures and genealogies to make sense of their work and to give it legitimacy for different audiences. They might place their work within a "globalisation of STS," displacing the centre of the study of science and technology from Europe and North America to reflect the changing global order (Dumoulin Kervran et al., 2018; Fischer, 2007). In this timeline, an objective expert, able to maintain distance from the object of inquiry, is fashioned in European enlightenment through a chasm between science and religion and the formalisation of scientific dis-ciplines (Daston and Galison, 2007). This happens just as new nation states emerge in Europe and impose imperial domination elsewhere. A second shift occurs around World War II, when military funding for sciences and establishment of scientific organisations like the National Science Founda-tion (NSF) frame the basic sciences as pure and curiosity-driven. By then, scientists were already called on by the state as experts to rationalise its actions (Sismondo, 2010). A third shift occurs in the 1980s with entry of private finance and non-profits into universities and research enterprises, establishing patent offices and intellectual property regimes to manage flows of scientific and technical knowledge (Fortun and Fortun, 2019). A parallel movement for environmental justice, open science, and software emerged in the 1970s, coalescing into what we today think of as citizen science.

STS scholars might question this timeline altogether by pointing that scholars in the Global South have been writing on science and technology for decades and not just at the moment of "globalisation of STS" (Subramaniam, 2019; Varughese, 2020). Since at least the 1970s, scholars have been commenting on the relationship of science with the state and the public. This scholarship is produced from Indian universities and research institutions like Jawaharlal Nehru University's (JNU) erstwhile Centre for Interaction of Science and Society in 1970, which was shut down following its critique of nuclear energy during Indira Gandhi's tenure as India's prime minister, and reopened as the Centre for Studies in Science Policy (CSSP) in 1996. The National Institute of Science, Technology and Development Studies (NISTADS) was established in 1973 under Government of India's Council for Scientific and Industrial Research (CSIR).

We begin this section with these institutions because our interlocutors grapple with their legacies in the here and now. They face dilemmas about when and where they should locate their environmental advocacy. They wonder how to build on legacies of environmentalism and public scholarship in India while responding to transnational conditions. The people we write about are located in institutions like the Centre for Science and Environment (CSE), CPR, the IITs, state and private universities, non-profits, and consultancies. They keep track of what's happening in conversations elsewhere and frequently travel to air pollution and environmental conferences around the world. They not only talk about India's and Delhi's air, but they also talk about the air in Beijing, Ulaanbaatar, Los Angeles, Berlin, Mexico City, and Nairobi.

How do we characterise this community of practice and the kind of thinking and public work it does? Instead of talking separately about experts and advocates, we talk about 'expert-advocates' to resist the genealogies that labelling someone as expert or advocate alone would signify. Our interlocutors are not just scientists or those trained professionally in technoscience, but have followed technoscience as it developed in India and elsewhere through different paths. They have learned to communicate and translate their expertise across silos. Therefore, the first step towards characterising these communities is to write about how legacies of environmentalism and public scholarship influence expert-advocates in the present. We borrow the microbiologist and historian of science Ludwik Fleck's ([1935] 1979) concept of "thought style" to talk about how expert-advocates communicate their knowledge.[3] Fleck argued that these communication events disrupt usual ways of knowing and doing as they call on the expert to adopt a distanced view of their own professionalisation and think about the relevance of what they do. We think with Fleck to write about the legacy of transnational conditions of doing science and technology. We find that expert-advocates contend with two different thought styles—*science for regulation* and *science for public understanding*. At their intersection, a new style is being worked out—a *science for advocacy*.

Science for regulation

In 2019, after expert-advocates had spent years talking about the necessity of a pan-India governance mechanism for air pollution, the Ministry of Environment, Forest and Climate Change (MoEFCC) circulated the National Clean Air Programme (NCAP) (MoEFFCC, 2019) after a round of public consultations. NCAP proposes an extensive urban and rural monitoring network, listing 102 cities which have consistently violated annual National Ambient Air Quality Standards (NAAQS), for taking priority action (further details in Chapter 4). It has 42 action points across different domains for improving air quality—vehicular transport, dust, biomass burning, industrial and construction sources. Two statements in the NCAP pique our interest:

> While developing and implementing technologies, it is of paramount importance that the technology suits the Indian scenario with respect to short- and long-term ecological and environment impacts, social infrastructure, cultural ethos, and characteristics of the Indian economy.
>
> (p30)

> While there is no denial on serious health implications, attributing one to one correlation and number of deaths due to air pollution needs to be further investigated and supported by indigenous studies. More authentic Indian data and studies may further strengthen our efforts and public participation in improving the air quality.
>
> (p51)

The call for more Indian data, technology, and studies is usually received as an irrational response to an already-existing global consensus (led by the World Health Organization and Health Effects Institute, for example) that air pollution undoubtedly kills. Do we really need more air quality data from various scales, of various pollutants, and exposure curves from various types of bodies and populations to tell us that people are dying at an alarming rate?

Expert-advocates find this ridiculous. Four comprehensive source apportionment studies exist for Delhi (Dubash and Guttikunda, 2018). Among them is the highly-cited IIT Kanpur report, after which dust control became a top policy priority (IIT Kanpur, 2016). These apportionment studies tell us what sectors and sources pollute the most, how much, and when. They agree that the air in the winters is worse. All point to high contribution of power plants, vehicular emissions, biomass/waste burning (including stubble), construction activities, and diesel generator sets, only varying by degrees in their estimations. A 2018 report published in *The Lancet* authored by 80 transnational collaborators from across disciplinary and institutional locations demonstrated beyond doubt the adverse health impacts of air pollution, even showing which states are affected more (the poorer ones, of course).

Yet, each time a central government minister makes some version of the "lack of evidence" argument to the media, expert-advocates respond with even more evidence while they continue to organise ever-larger advocacy networks. These collaborations look to produce, analyse, and rally colossal amounts of data to satisfy the "indigenous studies" benchmark. They include the "Atmospheric Pollution and Human Health in an Indian Megacity" programme, jointly funded by UK research agencies and the Indian Government. Enrolling 42 organisations with research teams led by over 100 scientists, this mega project will, over 4 years, tell policymakers what pollutes, how much it pollutes, where it pollutes, and who it pollutes. Other projects—some India-wide (Gordon et al., 2018), some focused on different Indian cities (Guttikunda et al., 2019)—are also underway to identify and quantify pollution and associated health effects. Documents published by these collaborations acknowledge that we need large-scale epidemiological studies that use a combination of monitoring, satellite, and personal exposure data for long-term health effects.

Delhi's expert-advocates are thus caught in a difficult position—they contend that enough is already known about the problem, but have to respond to the Indian Government's denials by organising themselves for new research. Their thought style is shaped by a global legacy of air pollution science that has established long-term, place-specific, cohort-based epidemiological studies as the *threshold* for arriving at air quality standards and air quality governance. The 'Harvard Six Cities Study' started in 1974 by a group of professors based at the Harvard School of Public Health sampled around 8,000 adults and 14,000 children controlled for smoking and age in six US cities for 16 years, publishing the results in 1993. It surmised that residents lost 2 years of their lives in cities heavily affected by air pollution, which amounted to massive losses in a large population

(Dockery et al., 1993). The study was met with ferocious contestation from industry, which alleged a lack of transparency because participants' clinical data were kept confidential (Fuller, 2019). The results were replicated in the 2000s and found to be consistent with even more locations. With the Harvard study, particulate matter became a pollutant of public concern (Grant, 2012).

These global conversations on particulate matter and environmental conferences like Stockholm, Rio, and Aarhus, influence air pollution governance in India. Even in the absence of the Six Cities type of study, air quality standards were notified for the first time in India in 1982, alongside other legal protections like the Wildlife Protection Act of 1972, Water Act of 1974, Air Act of 1981, and the Environment Protection Act of 1986. Unlike nuclear, space, and meteorological sciences, where security concerns have determined public involvement with science (Abraham, 1999; Phalkey, 2013; Dash, 2020), these concerns have not, until recently, been observed in the air pollution sciences. The NCAP, however, seeks to control air pollution science, arguing for centralisation of monitoring data and its processing (MoEFCC, 2019). It further advises the media to not report "perplexing statistics . . . to create an ambiguous public perception" (p56) and calls for sensitisation of media when reporting international studies.

When the goal of expansion of air pollution monitoring endeavours in India is read together with calls for indigenous studies, it points to a gap between lofty aims and a reality that does not seem to match. Observers frequently express frustration with the regulatory science conducted by the Central Pollution Control Board (CPCB) and State Pollution Control Boards (SPCBs), citing lack of infrastructural and resource capacities, the move towards centralisation and privatisation is argued to bring transparency, accountability, and better governance. It aims to position India as a self-sufficient, thriving nation that does need not rely on external expertise. The attempt towards this end in Indian science and technology is not new, seen previously in biotechnology (Sunder Rajan, 2006), telecommunications (Irani, 2015) and water provisioning (Anand, 2015).

With the calls for indigenous data, air quality monitoring is now becoming pivotal in regulatory science. Different types of monitoring strategies privilege different types of interventions, and accordingly, different institutional contexts and scientific disciplines. Ambient monitoring helps to make sense of urban air and privileges chemists, physicians, meteorologists, and atmospheric scientists. Satellite monitoring helps to make sense of regional and national air and privileges remote sensing and GIS-based sciences. Personal exposure monitoring helps to make sense of individual and aggregated bodies and the microenvironments they reside and move about in.

The faith in monitoring as the golden standard for air pollution regulation remains salient in all these strategies. Air quality data modelling is used to fill in the gaps when monitoring is not deemed sufficient, and even then, expert-advocates talk about more monitoring. When they talk about how much data is enough and good, they never talk about these different strategies separately. Everything is needed. In interdisciplinary science collaborations, different monitoring strategies can lead to differences over which type of data is perceived appropriate for policy (Garnett, 2017). For Delhi, we have noticed that expert-advocates call for more monitoring, arguing that other types of data complement monitoring data. Atmospheric scientist and air quality data modeller Sarath Guttikunda emphasized this in the first episode of the CPR seminar series on Delhi's air pollution as well. If the NCAP calls for more monitoring, it is *because* expert-advocates call for more monitoring. All our interlocutors mentioned the virtues of monitoring for air quality regulation, public outreach, and for advocacy.

As many research networks look to fill blindspots of monitoring, and millions of dollars are being spent to buy regulatory-grade monitors for urban and rural India, the conversation is shifting to how we do not know what to do with all the data being generated. Anthropologist Timothy Choy notes that knowing more about air does not necessarily produce clearer understanding, but "yields a sensation of incomplete knowledge, a vertiginous sense that there is always something in excess of the explanation" (Choy, 2010, p4). In keeping with Choy's observation, Anumita Roychowdury of CSE, an eminent advocate of air pollution interventions in Delhi since the 1990s, wonders (2020):

> Even after massive investments in 38 real-time monitoring stations . . . Delhi is still guessing if its pollution is rising or falling.

Real-time monitors that are supposed to account for transparency according to NCAP are not even considered for regulatory purposes, because they do not meet the legal requirements for compliance reporting (Somvanshi and Roychowdhury, 2020). While there is a very real danger that demand for more data for regulatory purposes would cause expert-advocates to run on a never-ending "data treadmill" (Shapiro et al., 2017), two shifts that we outline in this section remark on the possibility of air quality data for public action. First, the proposed centralisation and privatisation of regulatory air quality data with demand for indigenous epidemiological studies signals that contesting interpretations of air quality data will eventually arise. Second, the virtue associated with the act of collecting air quality data invites more people than before to engage in air pollution advocacy. Both these shifts become crucial for our expert-advocates when they mobilise an air pollution science for public understanding.

Science for public understanding

Expert-advocates make a distinction between data for different ends. They want to make the methods and results of regulatory air quality monitoring activities and source apportionment studies available for the public. They want to translate the results of international and national studies for wider dissemination. There is an expectation that with public outreach through the translation of science, enough public pressure will be created to make air pollution an electoral issue. Some think that this has already happened for Delhi, counting the list of visible air pollution interventions by the Aam Aadmi Party (AAP), such as the odd-even, which showed political commitment towards improving air quality (Ghosh and Harish, 2019).

The expectation that the translation of regulatory science creates an electoral demand for air quality is not without precedent either. In the 1990s, when changes to Delhi's fuel policy were being debated in the Supreme Court, government officials at various points articulated a lack of public pressure for not implementing extant regulatory standards (Sharma and Roychowdhury, 1996). Environmental advocates back then faced a similar challenge of amassing scientific evidence and translating it to the public. They were influenced by industrial disasters abroad, like Chernobyl, and those at home, like Bhopal and Shriram oleum gas leak, which changed environmental governance globally so that these disasters could be avoided elsewhere (Fortun, 2001). A style of environmental advocacy emerged during the late twentieth century in India, when engineers and business graduates realised the need to communicate the risks of chemical hazards and toxicities to the lay public. The organisations they set up were different from others organised around wildlife and forest protection or resistance to large infrastructure projects, in that their vision was to create community capacity for science. Three of these were Delhi-based: Toxics Link, Hazards Centre, and CSE. CSE was founded in 1980 by Anil Agarwal, a mechanical engineer from IIT Kanpur and a science journalist. He was joined by Sunita Narain, a student leader from Kalpvriksh. An early CSE communication was the *First Citizens' Report* in 1982, inspired by a brief "state of environment" report in Penang, Malaysia (Guha, 2002). Toxics Link was founded in 1994 by Ravi Agarwal, a communications engineer and MBA graduate. Its first report in 1998 was about food adulteration. The Hazards Centre was founded in 1997 by Dunu Roy, a chemical engineer from IIT Bombay, and published its first report on air pollution in Uttar Pradesh in 1997. Together, these organisations have published hundreds of reports and working papers using independent scientific research and drawing on research from elsewhere. The founders of these organisations have served on science and technology committees of various governments to advocate for a safer, healthier environment.

Of these three organisations, CSE emerged as the most prominent advocate around air pollution in the late 1990s and early 2000s, even though the other two continued to produce their own reports and pamphlets. CSE employed various strategies to that effect. Its anti-diesel campaigns called industrialist Rahul Bajaj "environmental criminal no. 1" and termed the Indian Council of Medical Research (ICMR) the "Indian Council for state-sponsored promotion of slow murder." It lobbied with politicians and bureaucrats to transform India's fuel policy and Delhi's transportation policy. Scientific debates raged between CSE, the judiciary, and governments, as documented by Awadhendhra Sharan (2014). Anil Agarwal joined the bureaucrat Bhure Lal with representatives from government and industry to form the Environmental Pollution (Prevention & Control) Authority in 1998 under the Supreme Court's directive to arrive at scientific consensus. What differentiated CSE from Toxics Link and Hazards Centre in our observations is that it pushed for judicial intervention through gathering technoscientific expertise within the Environment Pollution (Prevention and Control) Authority (EPCA). Whereas Hazards Centre continued to report about livelihoods in Delhi's physical and social peripheries and Toxics Link about chemical toxicities in the same areas, the CSE agenda at the time was to push for cleaner air through reforming transportation through the judiciary. It spoke little about the consequences of those interventions. Guha (2002) notes that as CSE grew, published, and campaigned, it maintained a distance from other campaigns of social and environmental justice at the time, especially the anti-dam Narmada Bachao Andolan (NBA). With help from judiciary and similarly inclined bureaucrats like Bhure Lal, CSE continued to contribute significantly to air pollution management through the EPCA, for instance, with the 2017 Comprehensive Action Plan for Air Pollution Control in Delhi and National Capital Region (NCR).

CSE works closely with industries to advocate for monitoring of industrial emissions, publishing manuals for installing and operating Continuous Emission Monitoring Systems (CEMS). One of us (Prerna) was present at the CSE-organised CEMS India Conference in 2017 held at the Leela Hotel, where the manual was released in collaboration with the Swedish Environment Protection Agency and the West Bengal Pollution Control Board. Representatives from non-profits, consultancies, governments, and industry spoke in turns about the necessity of monitoring industrial emissions, and an exhibition on the latest monitoring technologies, led by national and international companies, bustled with promise in an adjacent exhibition hall. This style of environmental advocacy, presenting primarily as a technoscientific matter of concern, has helped the CSE to advocate beyond political affiliations.

When expert-advocates today do science communications and outreach work, they contend with this style. The impulse remains to communicate scientific research, but with calls for indigenous studies by the Indian Government, mentioned earlier, the questions are now framed in terms of data that are easily accessible to the public. A large endeavour of expert-advocates is to communicate existing research. In March 2019, one of us (Prerna) attended a public workshop at IIT Delhi which taught non-experts how to process, analyse, and visualise air pollution data. Sarath Guttikunda has published primers on air pollution monitoring, source apportionments, and air quality management on his website "Urban Emissions" so that bureaucrats, policymakers, urban planners, activists, and other actors can learn technoscientific vocabularies and techniques. Pallavi Pant, of the Health Effects Institute (that now has ownership of the Harvard Six Cities Study data), maintains an extensive database of air pollution research and journalism on the blog "Air Quality in India," where she also interviews scholars, activists, and entrepreneurs associated with India's air pollution.

A related endeavour is to clean and present air quality data in attractive and easily understood formats. OpenAQ is a global platform where atmospheric scientists and engineers extract real-time government monitoring data and store it so that they become "historical data." These data can be accessed by anyone, though they are most of use to various app developers, allowing construction of further digital media. A popular example is Smokey, an air quality chat-bot in the form of a spectacled owl. Smokey replies instantaneously on Twitter and Facebook; needing only one's location, it presents the user with real-time data derived from OpenAQ, explains the pollutant of concern, and even offers links to buy products to protect oneself—and it does so in several languages. Another way to access monitoring data is to buy one of several sensors on the market. Indeed, using readily available low-cost sensors, many of Delhi's residents now monitor their breathing zones—their bedrooms, kitchens, cars, and officers—to guide decisions from exercising outdoors to using purifying technologies, as we further detail in Chapter 5. These sensors form alternative monitoring infrastructures which can contest government monitoring data, especially amidst suspicions of manipulation of air quality data during peak-pollution episodes, such as Diwali or the annual winter smog.

This style of framing environmental advocacy as questions of data accessibility and availability obviously excludes those who cannot interpret technoscientific data (mostly in English). But this framing has also resulted in an expansive notion of what that data could be and where it should come from. Take this quote from Pant:

> There is the science part, and then there is the other part of how we bring this data to public, how do we make it easily available . . . People can now buy sensors and see that the air pollution is really high and they can question the government . . . If I talk to people on [the] road about air pollution, they don't want to know about what model I used, how many variables are in that model, how you analysed the sample. What they want to know is how what they're doing is affecting their everyday live[s] and to embed air pollution in everyday experiences. If you remove the scientific terminology and start talking to people, a lot of people have noticed what causes air pollution and what are the problems.
>
> (Personal communication, 2017)

Pant points us to how low-cost sensors have become important to establish air pollution as a social concern. She intervenes in a popular narrative that residents of Delhi don't know and don't care about pollution by pointing to an absence of the *kind of data* that is missing from discussions: qualitative and textured data from people's everyday lives. Christa Hasenkopf, a founder of OpenAQ agrees:

> You can spend years developing the perfect data format. But our main thing was getting the data out there and see how people would use it. Because you cannot always anticipate how people will be using the system or meeting their data requirements . . . Air inequality is a term that our community came up with. We did a workshop in Sarajevo, Bosnia, a highly-polluted place in the winter, in terms of PM2.5. It was not data-based, but involved connecting people. They decided to write a community statement. And one of the participants in that workshop presented it to the Bosnian Parliament. It's not necessarily about the open data, but getting a group of people together to work nominally around the data in our platform and taking it into a completely different direction and still fight air inequality. To me, that was a win.
>
> (Personal communication, 2017)

Hasenkopf, like Pant, sees the promise of open data to make air pollution a public concern. She too uses an expansive notion of data to foster a community to advocate for air quality. Scholars have commented before on the limitations of enumerative practices for environmental advocacy, arguing that they lead to a "deficit model" that proceeds from assuming a lack of understanding on part of the non-expert public, and sets up knowledge transfer as the solution (Bickerstaff, 2004; Cupples, 2009; Dalborne and Galusky, 2011). We acknowledge these legacies of citizen science, including in India (Varughese, 2020). But responding to the call for contextualising citizen science research (Kimura and Kinchy, 2016), we think it is important to acknowledge that framing questions around data in the expansive way Pant and Hasenkopf do, allows for expanded deliberation with air pollution advocacy instead of it being centred around a few key figures or organisations. Both Pant and Hasenkopf do not see the public as lacking understanding, but are finding ways to connect existing information and capacities. We read these instances as "critical data practices" (Fortun et al., 2016), where advocates identify science for social concern and connect information that allows for new forms of advocacy to emerge.

Up to this point, we have noted two thought styles our expert-advocates contend with: science for regulation, and science for public understanding. Both are shaped by environmental advocacy and science at the global, national, and urban scales. Both are restricted and activated by the possibilities of air quality data. We now turn to the dilemmas expert-advocates face when these thought styles intersect.

Science for advocacy

As shown earlier, with the state's persistent denialism and the increasingly felt need to expand the domain of science beyond its narrowly defined epistemic community, there is the emergence of the expert-advocate. We now expand on the second part of this concept. Advocacy implies connecting with both the state and the public. Expert-advocates therefore contend with the following question: What kind of science, and data, are needed for advocacy? In this section, we call this new discursive space "science for advocacy." Our intervention is to show the dilemmas and caveats of this space that expert-advocates may keep in mind as they think through this question. We outline two dilemmas and their attendant caveats. First, contending with demands for science for regulation and science for public understanding produces a conflict of time. As expert-advocates acknowledge the need for long-term epidemiological studies for regulatory purposes, they also advocate for low-cost ambient and personal monitoring for public understanding of air pollution. Second, conversations on Delhi's air undoubtedly shift attention to air elsewhere, but the coordinates of science for advocacy stick stubbornly to Delhi.

Time for science, time for advocacy

Expert-advocates contend often that Delhi, and India, has very little time to clean its air. At one of CPR's seminars, Anumita Roychowdhury of CSE impressed this idea of time upon us: "We don't have 60 years, we have three." Phrases for interventions like low-hanging fruits, leapfrogging, real-time, and time-bound targets inform deliberations. Their descriptors for air are also temporal—air in the winters is worse, and annual air quality trends are better for regulatory purposes than daily air quality. CSE's report *The Leapfrog Factor* (Roychowdhury et al., 2006) asserts that air pollution in Asian cities cannot wait for science to catch up. What worked for Delhi in the 1990s was leapfrogging air quality and fuel standards. Expert-advocates wish to script a fresh success story of cleaning Delhi's air, built on the strongly held belief that the environment cannot be sacrificed at the altar of development.

In this temporal horizon of science for advocacy, organisations like CPR, CEEW, Harvard-IFMR's Evidence for Policy Design (EPoD) Delhi, Energy Policy Institute at the University of Chicago Center in Delhi (EPIC-India) organise their air pollution advocacy events in ways that ensure a year-round buzz on air. As we mentioned in this chapter's introduction, this was the intention of CPR's ten-episode seminar series. The dilemmas of time allow these organisations to advocate for specific policy interventions. EPIC's lead researcher Michael Greenstone's team built the Air Quality Life Index (AQLI), which aims to let people know when air quality is worst and when it is not advisable to go out and breathe. AQLI's website asks: "How much longer would you live if you breathed clean air?" When we zoom into the website's map for Delhi, it tells us that we could live a decade longer if Delhi's air met WHO guidelines for PM2.5. The framing of advocacy in terms of life expectancy maintains the concern around dying due to air pollution, which is a powerful framing, with the potential to impact the audience deeply. It also allows expert-advocates to call for policy interventions that respond to urgency with innovation. EPoD, along with EPIC and the Abdul Latif Jameel Poverty Action Lab (J-PAL), started the world's first emissions trading scheme for particulate pollution in the state of Gujarat.

As mentioned earlier, collaborations with industries and non-profits associated with private universities in the US point to a new direction in which science for advocacy is moving in response to frustration with older styles of advocacy and governance. The dilemma of time is resolved through the innovatory promise of market interventions, even though scholars have criticised cap-and-trade mechanisms for reinforcing existing inequalities. Working with industry and organisations is not new for science and technology in India, as we have mentioned earlier. The Happy Seeder, which we mentioned at the start of the chapter, was also developed as a non-profit-industrial collaboration. The Nature Conservancy, the International Maize and Wheat Improvement Centre (CIMMYT), the University of Minnesota, the Indian Council of Agricultural Research (ICAR), and the Borlaug Institute for South Asia (BISA) worked to create an alternative to crop residue burning for Indian farmers. But the combination of framing science for advocacy under life expectancy and market interventions pushes the conversation in a specific direction, which other groups and alliances find difficult to keep up with.

The story of the Public Health Foundation of India (PHFI) in these changing conditions speaks of the caveats we should keep in mind. Under the public–private partnership model in research, PHFI served as India's largest public health advocacy group and provided technical assistance to the Ministry of Health and Family Welfare (MoHFW) for its immunisation programmes. One of its concerted efforts was an anti-tobacco campaign, for which it was alleged to have used foreign funding from the Bill and Melinda Gates Foundation to lobby with politicians and media groups. This was considered a violation of the Foreign Contribution Regulation Act (FCRA) by the government, and PHFI's license to receive foreign funding was cancelled along with 20,000 other organisations like Greenpeace India (Sharma, 2017). Many transnational research collaborations were frozen immediately.

PHFI is one of the few organisations which framed air pollution explicitly as a matter of environmental justice. A report from the PHFI (2017) mentions that though air pollution is thought to affect all equally, vulnerability is, in reality, unevenly distributed. The class aspect is clear in their work: "urban upper-middle classes were better equipped in knowledge and resources to seek solutions to poor air quality, in comparison to urban poor classes" (Ibid., p40). It cites a social science study (Ramaswami et al., 2016) which surveyed three Delhi neighbourhoods with different socioeconomic profiles and interviewed waste handlers, showing that they were quite aware of health risks, but it was low on their priorities. It builds on a report by MoH and FW (2015) to argue that the urban poor are doubly burdened by ambient and household air pollution. Instead of outlawing certain practices, the report proposes affordable distribution of gas and electricity. Similarly, it does not blame rural households for not adopting improved cook-stoves (even though that is a decades-long intervention), offering alternative explanations for non-adoption. This is in sharp contrast with the influential IIT Kanpur report cited earlier that framed farmers as indulging in wasteful practices that ought to be banned outright (p278). The way emerging institutional arrangements that limit how expansively air quality data, and air pollution science, are to be understood are therefore issues that expert-advocates must think through.

Place matters

The second dilemma expert-advocates face is in *placing* science for advocacy. While we have written elsewhere about the questions of scale (Negi and Srigyan, forthcoming) around Delhi's air—framing it as simultaneously a national, regional (IG plains), and transboundary concern aided by different monitoring and governance strategies—we use this section to signal other shifts: the rise of the big urban sciences and a move towards talking about the mining districts of central India. The first shift simultaneously places Delhi within a geography of megacities troubled by air pollution and talks about smaller "tier II and III" cities which continue to violate air quality standards. The second shift is newer, a placeholder for talking about questions of political economy and environmental injustice presently missing from conversations on Delhi's air.

UrbanEmissions leads the Air Pollution Knowledge Assessment (APnA) city programme, where information about emissions, meteorology, and source-specific pollutant concentrations for about 60 cities has been published as a series of attractive pamphlets; the first cluster released in 2017 and the second in 2019. Its methodology involves making a massive publicly accessible database from disparate data sources to form an emissions inventory, creating a spatial model of air pollution using meteorological models for climate variables and dispersion models for concentration, and provides a mechanism for air quality management according to sources and seasons (Guttikunda et al., 2019). The results and recommendations from this enterprise are disseminated as easy-to-read pamphlets where clusters of towns make "airsheds" (see Chapter 4). In towns where ambient air quality monitoring has not yet reached or reached only sporadically, APnA is an inspirational example of a critical data practice, bridging the gap between the "different airs of air pollution science" (Garnett, 2020). Guttikunda's mobility in advocacy circles ensures that each of the 60 APnA case studies also circulates.

We discuss in the following chapters about how the rise of big data under the framework for smart cities has worked to advocate for airshed-based style of air pollution governance, but we note here that one effect is to see the urban as a space for experimentation. This is tied to the success stories told by expert-advocates of earlier interventions to clean Delhi's air and to the odd-even "experiments" undertaken by AAP government in 2016, but differs in that the scale is not Delhi alone but the *urban* with its attendant demography and complexity. Scholars have written about urban living laboratories (ULLs) in Europe serving as civic platforms to facilitate sustainability transitions and promising openness actually have ambiguous relationships with both (Bulkeley et al., 2018). As in the case of India's hackathon spaces (Irani, 2015), experimentation, rather than translating results of experiments into environmental education, remains the primary motive. During the COVID-19 pandemic lockdown in 2020, particulate matter dropped by around 50 per cent as transportation and industry halted in Delhi (Mahato et al., 2020), and we observed a flurry of activity on social media calling the lockdown an experiment in improving air quality (discussed in greater detail in the Postscript). Meanwhile, a migrant exodus from Delhi was underway towards those tier II and III cities, and into the rural areas of India.

The air quality in places where these migrants were relocating to is not held by expert-advocates to be much better either. The NCAP in fact advocates for establishing a rural air quality network. In our conversations with expert-advocates and our participation in events organised by them, the rural areas of India and the mining districts of central India emerge as new centres where science for advocacy should pay attention to. Ronak Sutaria, founder of Respirer Living Sciences and a participant in conversations around urban sciences, says:

> We are working in Singrauli and Korba. If you go to the Atmos [discussed in Chapter 5] map, you will see some monitoring there. These are our very critically polluted areas. We are working with absolutely on-the-ground organisations there. They don't have a high-profile website or visibility, but they do a lot of work on the ground.
>
> (Personal communication, 2017)

Sundeep Kumar, coordinator of the Healthy City Alliance (HCA),[4] an umbrella organisation that has recently emerged to tackle air pollution advocacy at various scales, discusses the Alliance's work beyond Delhi:

> The air in Singrauli, Korba, Dhanbad is also polluted. If we did not have people or organisations who are already working there, I would seek people doing similar work and get them to be a part of [the] collective . . . There might be ideas coming out from Singrauli or Korba. On a meta-level, poor neighbourhoods in Delhi would be Korba from an India point-of view. It expands the base because earlier you would only talk about people who are working in the mines. But now you're talking about someone who's living in Raipur, which is far away from Korba.
>
> (Personal communication, 2018)

We welcome the move towards shifting the coordinates of conversation towards these areas, but the styles and dilemmas of advocacy we have written about so far make us cautious. Both Sutaria and Kumar acknowledge working with local organisations, but due to existing distance from allying itself explicitly with questions of environmental and social justice and changing conditions of doing science and technology in India, how would existing grassroots advocacy in these areas transform? Further, while Korba and Singrauli allow for a critique of state- and corporate-led enterprises that have caused adverse health effects, they cannot stand-in for similar analysis of poorer neighbourhoods of Delhi, especially as these neighbourhoods have been partly created by earlier interventions to push industries and polluting facilities to Delhi's peripheries. We find that though science for advocacy in Delhi has shifted attention to other scales that allow for critique of political economy, it so far does not engage enough with peripheral spaces and livelihoods of Delhi itself.

Conclusion

Air as a fractured condition for breathing and existence sometimes causes advocates and scholars to imagine a future where air would be once again a shared condition of life. Our reading of science for advocacy around Delhi's air takes us instead in an opposite direction. The dilemmas of time and place that characterise science for advocacy craft a distance from critique of political economy and justice in Delhi, even as they facilitate these conversations beyond Delhi. We call for an analysis of new institutional contexts that take advantage of these dilemmas, such as how changing conditions of science and technology have made it difficult to articulate science for advocacy along explicit political lines. Our interlocutors, for instance, would do well to bring into conversation other expert-advocates who have, for long, taken seriously the questions of livelihoods and residents of Delhi's peripheries that earlier styles of advocacy did not sufficiently engage with. Science for advocacy calls for a sustained engagement with existing movements for environmental and social justice. How might such engagements be facilitated?

We elaborate on two tactics in our Postscript—slowness and carefulness—that air pollution advocacy in general could benefit from. We now signal those tactics here. When Hanjra spoke in the CPR seminar room, it opened a collaborative moment. His embodied practice of working and experimenting in the farm collided with Jat's understanding of agriculture and productivity. But friction need not be unproductive. Both had invested years in their research and came from locations beyond Delhi to intervene in conversations on Delhi's air pollution. Both had experienced uncertainties and failures in their science and advocacy, as well as successes beyond their personal spaces. Working at that collaborative moment means talking frankly about the relations of political economy between them that made it possible for them to be seated together in CPR's seminar room. It means asking questions that would lead either party to acknowledge uncertainty and partial knowledge, rather than placing them as antagonistic actors. Being careful partly means creating space for these older questions of political economy: of class, caste, and gender.

Notes

1 The recordings of the seminar series are available at the website for the Centre for Policy Research.
2 Kim Fortun (2001) writes about the contradictory obligations that advocates experienced after the 1984 Bhopal gas disaster, calling communities of people who come together (or not) within a specific moment to speak on the disaster's aftermath as "enunciatory communities." They are different from stakeholders, as they might not agree what is an appropriate matter for concern. They are also different from epistemic communities, because they are less bound by disciplinary allegiances or faithfulness to one form of expertise. The heterogeneous communities we describe in this chapter are similar to Fortun's enunciatory communities, but are different because they form in response to slower disasters (air pollution in our case) and require new ways of responding in time.
3 As cited in Löwy (2016), Fleck defined thought style as "a set of findings meanders through the community; becoming polished, transformed, reinforced or attenuated, while influencing other findings, concept formation, opinions and habits of thought."
4 Both the person's and organisation's names have been anonymised.

References

Abraham, I. (1999) *The making of the Indian atomic bomb: science, secrecy and the postcolonial state*. Hyderabad: Orient Longman.

Anand, N. (2015) 'Leaky states: water audits, ignorance, and the politics of infrastructure', *Public Culture*, 27(2), p305–330.

Baviskar, A. (2003) 'Between violence and desire: space, power, and identity in the making of metropolitan Delhi', *International Social Science Journal*, 55(175), p89–98.

Bickerstaff, K. (2004) 'Risk perception research: socio-cultural perspectives on the public experience of air pollution', *Environment International*, 30(6), p827–840.

Bulkeley, H. et al. (2018) 'Urban living laboratories: conducting the experimental city?', *European Urban and Regional Studies*, 26(4), p317–335.

Choy, T. (2010) *Air's substantiations*. Paper for Berkeley Environmental Politics Colloquium. Available at http://globetrotter.berkeley.edu/bweb/colloquium/papers/ChoyAirEP.pdf (Accessed 9 September 2019).

Cupples, J. (2009) 'Culture, nature and particulate matter: hybrid reframings in air pollution scholarship', *Atmospheric Environment*, 43(1), p207–217.

Dalborne, J. and Galusky, W. (2011) 'Toxic transformations: constructing online audiences for environmental justice', in Ottinger, G. and Cohen, B. (eds.) *Technoscience and environmental justice: expert cultures in a grassroots movement.* Cambridge, MA: The MIT Press, p63–92.

Dash, B. (2020) 'Science, state and meteorology in India', *Dialogue: Science, Scientists, and Society.* Available at www.dialogue.ias.ac.in/article/21688/science-state-and-meteorology-in-india (Accessed 18 August 2020).

Daston, L. and Galison, P. (2007) *Objectivity.* New York: Zone.

Dockery, D. W. et al. (1993) 'An association between air pollution and mortality in six U.S. cities', *New England Journal of Medicine*, 329, p1753–1759.

Dubash, N. and Guttikunda, S. (2018) 'Delhi has a complex air pollution problem', *Hindustan Times*, 22 December. Available at www.hindustantimes.com/india-news/delhi-has-a-complex-air-pollution-problem/story-xtLhB9xzNYeRPp0KB-f9WGO.html (Accessed 2 March 2020).

Dumoulin Kervran, D., Kleiche-Dray, M. and Quet, M. (2018) 'Going South: how STS could think science in and with the South?', *Tapuya: Latin American Science, Technology and Society*, 1(1), p280–305.

Fischer, M. J. (2007) 'Four genealogies for a recombinant anthropology of science and technology', *Cultural Anthropology*, 22(4), p539–615.

Fleck, L. ([1935] 1979) *Genesis and development of a scientific fact.* Chicago: The University of Chicago Press.

Fortun, K. (2001) *Advocacy after Bhopal: environmentalism, disaster, new global orders.* Chicago: University of Chicago Press.

Fortun, K. et al. (2016). Pushback: Critical Data Designers and Pollution Politics. *Big Data & Society*, 3(2).

Fortun, M. and Fortun, K. (2019) 'Anthropologies of the sciences: thinking across strata', in McClancy, J. (ed.) *Exotic no more: anthropology for the contemporary world.* Chicago: University of Chicago Press, p241–263.

Fuller, G. (2019) 'How one air pollution study from 1993 changed history', *Salon*, 24 May. Available at www.salon.com/2019/03/24/how-one-air-pollution-study-from-1993-changed-history/ (Accessed 3 July 2020).

Garnett, E. (2017) 'Air pollution in the making: multiplicity and difference in interdisciplinary data practices', *Science, Technology, & Human Values*, 42(5), p901–924.

Garnett, E. (2020) 'Breathing spaces: modelling exposure in air pollution science', *Body & Society*, 26(2), p55–78.

Ghosh, S. and Harish, S. (2019) 'Pollution is now politically salient in national capital', *Hindustan Times*, 14 September.

Gordon, T. et al. (2018) 'Air pollution health research priorities for India: perspectives of the Indo-U.S. communities of researchers', *Environment International*, 119, p100–108.

Grant, E. A. (2012) 'Prevailing winds', *Harvard Public Health: Fall 2012*. Available at www.hsph.harvard.edu/news/magazine/f12-six-cities-environmental-health-air-pollution/ (Accessed 18 August 2020).

Guha, R. (2002) 'Anil Agarwal and the environmentalism of the poor', *Capitalism Nature Socialism*, 13(3), p147–155.

Guttikunda, S., Nishadh, K. A. and Jawahar, P. (2019) 'Air pollution knowledge assessments (APnA) for 20 Indian cities', *Urban Climate*, 27, p124–141.

IIT Kanpur (2016) *Comprehensive study on air pollution and green house gases (GHGs) in Delhi*. Kanpur: Indian Institute of Technology.

Irani, L. (2015) 'Hackathons and the making of entrepreneurial citizenship', *Science, Technology, & Human Values*, 40(5), p799–824.

Kimura, A. and Kinchy, A. (2016) 'Citizen science: probing the virtues and contexts of participatory research', *Engaging Science, Technology, and Society*, 2, p331–361.

Löwy, I. (2016) 'Fleck the public health expert: medical facts, thought collectives, and the scientist's responsibility', *Science, Technology, & Human Values*, 41(3), p509–533.

Mahato, S., Pal, S. and Ghosh, K. G. (2020) 'Effect of lockdown amid COVID-19 pandemic on air quality of the megacity Delhi, India', *Science of the Total Environment*, 730, p1–23.

Ministry of Environment, Forests and Climate Change (2019) *National clean air programme*. Available at http://moef.gov.in/wp-content/uploads/2019/05/NCAP_Report.pdf (Accessed 2 August 2020).

Ministry of Health and Family Welfare (2015) *Report of the steering committee on air pollution and health related issues*. New Delhi: MoHFW.

Negi, R. and Srigyan, P. (forthcoming) 'Peopling technoscience: the sciences and publics of air pollution in Delhi', *Dialogue: Science, Scientists and Society*.

Phalkey, J. (2013) *Atomic state: big science in twentieth-century India*. Ranikhet: Permanent Black.

Public Health Foundation of India and Centre for Environmental Health (2017) *Air pollution and health in India*. Available at www.ceh.org.in/wp-content/uploads/2017/10/Air-Pollution-and-Health-in-India.pdf (Accessed 19 July 2020).

Rajan, K. S. (2006) *Biocapital: the constitution of postgenomic life*. Durham, NC: Duke University Press.

Ramaswami, A., Baidwan, N. K. and Nagpure, A. S. (2016) 'Exploring social and infrastructural factors affecting open burning of municipal solid waste (MSW) in Indian cities', *Waste Management & Research*, 34(11), p1164–1172.

Roychowdhury, A. (2020) 'How do we know air quality is changing?', *Down to Earth*. Available at www.downtoearth.org.in/blog/air/how-do-we-know-air-quality-is-changing-69268 (Accessed 18 August 2020).

Roychowdhury, A., Chattopadhyaya, V., Shah, C. and Chandola, P. (2006) *The leapfrog factor: clearing the air in Asian cities*. New Delhi: Centre for Science and Environment.

Shapiro, N., Zakariya, N. and Roberts, J. (2017) 'A wary alliance: from enumerating the environment to inviting apprehension', *Engaging Science, Technology, and Society*, 3, p575–575.

Sharan, A. (2014) *In the city, out of place: nuisance, pollution and dwelling in Delhi c. 1850–2000*. New Delhi: Oxford University Press.

Sharma, A. and Roychowdhury, A. (1996) *Slow murder: the deadly story of vehicular pollution in India*. New Delhi: Centre for Science and Environment.

Sharma, D. C. (2017) 'Concern over India's move to cut funds for PHFI', *The Lancet*, 389(1008), p1784.

Sismondo, S. (2010) *An introduction to science and technology studies* (2nd ed.). Chichester: John Wiley and Sons.

Somvanshi, A. and Roychowdhury, A. (2020) *Breathing space: how to track and report air pollution under the national clean air programme*. New Delhi: Centre for Science and Environment.

Subramaniam, B. (2019) *Holy science: the biopolitics of Hindu nationalism*. Seattle: University of Washington Press.

Varughese, S. S. (2020) 'Expertise at the "deliberative turn": multiple publics and the social distribution of technoscientific expertise', *Dialogue: Science, Scientists, and Society*. Available at http://dialogue.ias.ac.in/article/21998/21998-expertise-at-the-deliberative-turn-multiple-publics-and-the-social-distribution-of-techno-scientific-expertise (Accessed 18 August 2020).

Véron, R. (2006) 'Remaking urban environments: the political ecology of air pollution in Delhi', *Environment and Planning A*, 38(11), p2093–2109.

4 Governance and atmospheric citizenship

I wake up one early-winter morning, and, as is the case these days, pick up my phone. Usually, I check email or Twitter first thing. But it is less than a week after Diwali, and the entire North Indian region is enveloped by thick smog. I open an air quality app on the phone. It shows that the Air Quality Index across most of the city is over 400. That's severe. An hour later, I'm on the road listening to the radio. One of the city's most recognised radio jockeys is beside himself. "Why are we in this mess year after year?" he asks, "Why do we let the health emergency appear every winter?" Later that morning, I read an acquaintance's tweet on the subject: "Funny that people are having a go at Kejriwal for Delhi's poor air. He's doing all he can (some stuff being too populist for my liking) but he can't do jack about the main issue—stubble burning happens in Punjab and Haryana. And their CMs don't f***ing care. #DelhiChokes." He adds another message on how the Odd-Even scheme is "largely for optics" since vehicular pollution is not the main cause of pollution in Delhi. I take a quick look at a website that models air pollution by source, and reply: "right now, vehicles are to account for 15–20% of the pollution; exposure for some will reduce due to faster travel times, but it's not about Delhi as you rightly put it. Except the BJP propaganda machine is at it—just saw a WhatsApp forward from them." That meme, forwarded to me by an urban planning graduate no less, is on point. A stock image of Delhi Chief Minister Arvind Kejriwal coughing is accompanied by the following message: "*AAP sarkar ki kamayabi, pehle akela khansta tha, aaj poori Dilli khaans rahi hai*" [AAP government's success—he used to cough alone earlier, now entire Delhi coughs along].[1]

Living in Delhi is to be entangled with toxicity. The simple act of breathing in the conurbation is hazardous, and it is much worse for those who labour, cook with biofuels, and sleep in the open. As the note mentioned earlier indicates, air envelops not only our bare lives but also our political lives: opposition parties blame the government, protests call on the state to act, and election manifestos promise to clean the air. It is in this backdrop that this chapter attends to the governance of and civic engagement with air in the Delhi region. It takes up the ways in which these processes come to be, the shifting nature of popular claims for clean air, the question of who is left out from these spaces, and how the recognition of exclusion propels civil society reflection and recalibration, pushing them towards the two critical moves of translation and collaboration, even as the contours of an emergent "atmospheric citizenship" are being worked out in contemporary Delhi.

The United Nations Environment Programme (UNEP) defines environmental governance as the "rules, practices, policies and institutions that shape how humans interact with the environment."[2] In their review of environmental governance literature, Armitage et al. (2012) show how the concept has been broadened to encompass a range of institutions, norms, and practices. Scholars view environmental governance as inclusive of formal and informal rules, regulatory processes, conflict-solving mechanisms, and agents and networks at different spatial scales working broadly to "prevent, mitigate and adapt to global and local environmental change." (Armitage et al., 2012) While early work was focused on the state and the legal-administrative contexts within which environmental change was seen to take place, scholars like Erik Swyngedouw (2005) noted the rise of "governance-beyond-the-state," comprising of "horizontal associational networks of private (market), civil society (usually NGO), and state actors" operating within the larger context of "the rise of a neo-liberal governmental rationality and the transformation of the technologies of government" (p1991). In his work on political-economic shifts in Africa, anthropologist James Ferguson (2006) similarly described the manner in which transnational civil society groups stepped into the governance vacuum left by the neoliberal retrenchment of the state, thereby producing an institutional terrain largely devoid of accountability from below while pivoting on travelling transnational expertise. It follows that with air pollution too, there is an entire complex and path-dependent arrangement of public, private, and civil society organisations that shape the conditions within which differently situated populations breathe air. There has been a significant expansion in the number and influence of non-state agents from before when state-run scientific and regulatory institutions had a monopoly over scientific knowledge production and authority to act on concerns. In this chapter, we take up these concerns and processes. We begin by showing how two elements of governance, that is, it's regional dimension and judicial activism, have been important to the debate, and then outline the contours of an atmospheric citizenship, which allows air to be considered as a shared but differentiated condition, and around which claims are being made for healthier environments.

Governing air

Air kisi department mei hai hi nahi! Air ka koi department hi nahi hai! Air pollution to kisi ka hai hi nahi [Air is not within any one department! There is no department for air! Air pollution is not with anyone][3]

Attending to diverse institutional architectures around air pollution backs Follman's finding that urban environmental governance in India is characterised by a "multiplicity of agencies, overlapping jurisdictions as well as fragmented and ill-defined responsibilities" (2016, p2). At the central level, the MoEFCC is the umbrella ministry that oversees environmental protection through policy and regulatory means. The legal architecture of pollution regulation is laid down by the Air (Prevention and Control of Pollution) Act of 1981, which empowers the CPCB, aided by the SPSBs at the provincial level. The CPCB is tasked with monitoring pollution, its prevention, and enforcing standards for air quality. These agencies are, however, not empowered to impose penalties on offending parties but may only "direct the closure of an offending unit or cut off/regulate its water or power supply" (Ghosh, 2015, p13). That authority lies with the criminal courts.

Since air pollution governance involves a phalanx of different institutions—industry, transportation, land-use planning, etc.—it is no surprise that several other government departments have a stake in the issue, and their decisions have implications on it. The issue is of interest to the MoHFW, which has initiated research and action focused on the health impacts of pollution, bringing together epidemiologists, public health experts, and medical researchers. These two blocs—monitoring and regulation on the one hand, and exposure and health on the other—are largely disparate and their relations are characterised by rare interactions (Negi and Srigyan, forthcoming). Speaking of our region of interest, there is the Delhi Government, which has its own Department of Environment, whose Secretary chairs the Delhi Pollution Control Committee. At an even more fine-grained scale, municipalities and local governments are responsible for actions like waste management and road cleaning, which have direct implications on air pollution. In a city like Delhi, these areas are distributed across multiple authorities, with their own respective systems, logics, and election cycles.

An audit of the regulatory system noted the paucity of resources, absence of rewards and recognition, as well as lack of motivation within and coordination between agencies (IIM Lucknow, 2010). It is then not surprising that the boards look to limit prosecution, given the protracted nature of proceedings and their resource-intensive nature (Ghosh, 2015, p14). Similarly, poor coordination not only means that air pollution falls through the cracks—as the opening quote suggests—but also that a lot of routine work is replicated, wasting precious time and resources. According to Siddharth Singh, energy scholar and author of *The Great Smog of India*, in 2015, after a major public outcry, the Delhi Government commissioned a source apportionment study that cost a lot. Singh asks:

> What was the need for the study when . . . on the CPCB website, for Delhi's pollution, already a 600-page report existed? All you needed to do [was] literally google "Delhi and air pollution" and the first result on the top you click it, and that goes to [this report] which has all the sources listed.
>
> (Personal communication, 2019)

According to another researcher who works on the epidemiology of air pollution:

> Departments don't talk to each other. The Ministry of Health doesn't know what Ministry of Urban [Development] is doing. But these things are all related. The Ministry of Health should know what [the] Ministry of Agriculture is doing. Everyone should know everyone else's work, so that overall development and knowledge increase.[4]

While the institutional architecture has, over time, been tweaked, modified, and recalibrated as it encounters new questions and challenges, the arrangements do not seem to have been thought through comprehensively with long-term concerns and the urgency of coordination in mind. Some of these issues seem to have informed the formation of the new Commission for Air Quality Management (CAQM) for the NCR and adjoining areas via an ordinance in October 2020. We return to the commission later. In short, the "administrative tournament," as Singh (2018) calls it, leads to "delay and inefficiencies in implementing plans that involve action from multiple institutions" (p180). A way forward might be to create a larger campaign or organise a movement such as the *Swachh Bharat*, which itself is focused on waste management, around which different agencies reflect on their own relation with and actions around air pollution. Karnad (2020) puts it eloquently: "To truly revitalise our air, we need to change how we cook, build, farm, travel, consume, and produce—bearing in mind, through it all, how we breathe."

Governance is also impacted by shifts in the larger political economy within which pollution is being debated and acted on. Even before the COVID-19 pandemic, the millennial Indian growth story had been arrested, first, by the global crisis since 2008, and then the lasting consequences of demonetisation. In urban imaginaries, the erstwhile fantastic dreams of becoming the next Singapore via large-scale infrastructures have been largely replaced by mundane concerns like neighbourhood clinics, access to drinking water, and affordable electricity. In Delhi, this is linked to the rise of the AAP[5] and the shift in urban politics has not escaped the local governance of air pollution. AAP is visibly involved in Delhi's air pollution debate. One of its interventions is the vehicular rationing system called 'odd-even' (cars plying alternate days according to their number plates) was hailed as both radical (first of its kind in India) and progressive (attacking middle-class demand for more vehicles). The AAP government regularly mounts other visible anti-pollution campaigns, such as planting trees, using vacuum road-dust cleaners, and switching off vehicle engines at traffic signals. AAP is also represented in civil society discussions around air, as we repeatedly observed. It interacts with environmental advocacy groups in contrast to the previous era, when the civil society had to lean on the judiciary and popular media to jolt the state into action. Still, the mechanics of governance involve spaces and agencies beyond Delhi and its government, while the judiciary continues to remain an important locus of debate and action.

Regionalising governance

Scholars have long argued that polluted air is a transboundary concern, and the atmosphere, an "international commons" (Soroos, 1991). Geography and meteorology make it impossible to contain pollution in a bounded space or to prevent toxic air from entering particular regions from afar. Governing air, therefore, requires a regional imagination and spatially flexible instruments. Many scientists and advocates we spoke to repeatedly highlighted this issue, and regretted the disproportionate emphasis on Delhi in the media and public discourse on air pollution. They were keen to point out that the IG region and other places in the country, especially with a concentration of mining activities, were faced with an equal if not worse pollution problem (see Chapter 3). Delhi, it has also been argued, is affected significantly by sources of toxicity located elsewhere. It has been shown that external sources are responsible for anywhere between 20 and 50 per cent of the load, depending on the specific pollutant, with ozone accounting for the largest external contribution (Marrapu et al., 2014, p10629). This goes to show not only that simply relocating certain pollution sources from Delhi to its fringes is inadequate, but that regional environmental planning is required for comprehensive action. Alongside thermal power plants, a critical source of the city's pollution is the burning of farm residue chiefly in Punjab, Haryana, and Uttar Pradesh, which spreads all across the IG belt during early winter. The temporality and economics of paddy cultivation have together contributed to the scale of the problem (Kumar, 2017), and despite a 2015 ban by the NGT and numerous government initiatives, harvested fields continue to be set on fire each year. The problem is compounded by the lack of meaningful interstate conversation or coordinated response, and a high level of mistrust. While acknowledging the pollution caused by agricultural fires, for instance, the Punjab Chief Minister recently called his Delhi counterpart a "shameless liar" for blaming his state alone, adding that "air pollution in the national capital [was] directly related to the rampant construction activity, widespread industrialisation and total mismanagement of the city traffic" (Indian Express, 2019), in other words, causes internal to the city.

More broadly, the growing interest in the regional dimensions of pollution has led organisations like Greenpeace India to reframe air pollution as a spatially diffuse issue, terming their campaign "Clean Air *Nation.*" Similarly, Urban Emissions' APnA places detailed information on pollutant concentrations for 20 cities in the public domain. The Supreme Court-mandated EPCA—now defunct after the creation of the CQAM—was tasked with working on pollution in the entire National Capital Region, rather than the city of Delhi alone, as was the case originally. The central government has also built on such an understanding of the issue to launch the NCAP in 2019. The NCAP states that "since air pollution is not a localised phenomenon, the effect is felt in cities and towns far away from the source, thus creating the need for regional-level initiatives through interstate and intercity coordination in addition to multi-sectoral synchronization." (MoEFCC, 2019, p3) It aims to meet the air quality standards across different parts of the country through a 20–30 per cent reduction in PM2.5 and PM10 concentrations by 2024 relative to base year (i.e., 2017). To this end, it aims to strengthen monitoring capacities in medium and small towns and cities, which hitherto have been poorly served by the available monitoring infrastructures. In the next chapter, we show how this push has created the space for entrepreneurship in the low-cost monitoring sector. Here, we wish to note that NCAP envisages partnering with a wide array of non-governmental institutions to generate knowledge of air pollution—such as a National Emissions Inventory—to plan interventions and to implement them effectively. The very first point in the NCAP section on its approach, that is, its strategy to implement the plan, mentions "collaborative, multi-scale and cross-sectoral coordination between the relevant central ministries, state governments and local bodies" (Ibid., p24). It is, however, less clear on the mechanics of such an effort. In this scenario, two strategies are being discussed by engaged scholars and practitioners.

The first argues for a new, specialised, and empowered authority. These are desires to create the office of an "air-quality manager" (Bolia and Khare,

2018), preferably reporting directly to the Prime Minister. One of the main functions of this authority, it is argued, should be to oversee and coordinate efforts of involved agencies; acting in some ways like the Administrator of the US Environmental Protection Agency (EPA). Arguing that over 40 per cent of India's air pollution crosses state boundaries, the authors of a recent study (Du et al., 2020) call for a "more centralized form of environmental federalism." Relatedly, a recent report by the Collaborative Clean Air Policy Centre (CCAPC) proposes an "Air Quality Management District" for Delhi and NCR to promote airshed-level governance of pollution. Its role "will be to understand and take into account all the major sources in the airshed and develop air quality management plans for the whole region, rather than just for the individual cities falling in the region" (Khanna and Sharma, 2020, p10). The document cites the cases of the Air District in California and regional initiatives in China in the Yangtze and Pearl River Delta, respectively, to call for a similar strategy for the NCR. The idea is not entirely new. The Director of the Delhi Government's Environment Department had flagged the need for airshed management of pollution in the NCR at a 2017 meeting organised by the International Institute for Energy Conservation. In our reading, the newly promulgated commission (CAQM) tries to do a bit of both the aspects of regional governance we note: it is in part about airshed management, and in part about better coordination. The agency has been given dedicated staff and resources—which was always the limitation with the EPCA—and members are drawn from the various states of the NCR plus Punjab (presumably because of the crop residue issue), representatives from several central government ministries, and from the civil society. While it promises better coordination for effective enforcement, there is little reason to doubt that the balance of power in the CAQM rests squarely with the central government in terms of both the overall number of members and in the appointment of the permanent members. Placing environmental governance within its larger context, CAQM fits firmly within the shifts in Indian political geography towards greater centralisation (and authoritarianism), and so, fears that environmental management may become the pretext to further weaken the powers of the states must be considered sympathetically, and might actually hurt efforts at smoother coordination. As we noted in the previous chapter, environmental concerns cannot be discussed in a vacuum, and expert-advocates, therefore, have to be more serious about the political implications of their seemingly scientific/technical arguments, such as the one by the group cited earlier, which made a strong, science-based, case for centralisation of air's governance.

The second strategy towards regional coordination does not advocate a separate institution above the present architecture, but focuses on horizontal coordination, drawing the various involved agencies into sustained conversation aimed at directed actions. The key idea with such political work is trust, which, as we argued in Chapter 2, is not presumed to exist but emerges alongside collaboration. One nascent effort towards a horizontal approach is by a group of Indian parliamentarians who, according to member of parliament Gaurav Gogoi, "have put together a bipartisan group of MPs. [They] are trying to become influencers and encourage each other to work on different aspects of air quality management."[6] Gogoi adds that the group tries to educate itself in the technical nuances of pollution, while working to create a constituency within different political parties to push for coordinated air pollution action. This group would do well to study the experiences of Europe's Convention on Long-Range Transboundary Air Pollution (CLRTAP), in operation since 1979, which brought together the Eastern and Western regimes of the continent. If Cold War adversaries could converse, agree on policy, and coordinate action around air pollution, then surely Indian states and parties can too. The Convention resulted from the shared scientific realisation that atmospheric pollutants travelled hundreds of miles to cause damage in places far removed from their source. CLRTAP has separate protocols related to ozone, heavy metals, sulphur emissions, VOCs, and nitrogen oxides, each focused on the specificities of monitoring, reduction, and prevention of pollution through legally binding provisions. In an analysis of the convention, Lidskog and Sundqvist (2002) describe the careful manoeuvring that was required to bring together scientific best practices and governmental action in a transboundary framework. They point to the importance of setting up communication channels between the constituent units, of starting with relatively modest achievable goals which are progressively scaled, and of the negotiations through which science and policy are co-produced, rather than one or the other assuming pre-eminence in determining the trajectory of plans. In other words, the malleability of both politics and science was viewed as a productive point of exchange: politicians thought like scientists, and scientists, like politicians. The convention shows that tricky transboundary or interstate issues require deft manoeuvring as much as they demand technical capacities and evermore sophisticated knowledge. Yet, institutional interventions that hinge on managerialism rarely appreciate the politics of environmental governance (Negi, 2019), and therefore hardly deliberate on the political skills needed to build effective collaborations. As air pollution governance is regionalised around Delhi, more such conversations are urgently required.

Judicial interventions

A critical aspect of atmospheric governance in the Delhi region has been the role of the judiciary in adjudicating contentious issues, while forcing the state to act on these concerns. The roots of the present judicial debate on air pollution can, in some ways, be traced back to December 1985, when an oleum gas leak from Shriram Industries' chemical plant in the Moti Nagar area of Delhi killed one person and made many others sick. Environmental lawyer MC Mehta's writ petition in the Supreme Court for the closure of hazardous industries like Shriram, which operated in proximity to dense residential areas, was already pending when the incident took place. Exactly a year since the Bhopal disaster, the court was more alert than usual and took serious cognisance of the matter. It imposed strict regulations on the factory in addition to ordering it to create a separate fund to compensate those affected by the leak. The court also "moved from the specifics of the S[h]riram case to more general propositions regarding hazardous industries in urban contexts" (Sharan, 2014, p234). With it, air pollution became an important matter of concern to the court, leading to several orders that had enormous bearing on Delhi's air—and much more—in the following years.

In 1990, the freshly notified Delhi Master Plan 2001 stipulated that all hazardous and noxious industries were to be shifted out of Delhi. Like many other things contained in the nominally statutory Plan, this too wasn't seriously pursued by implementing agencies. But the provision was cited by the courts adjudicating on air pollution to order the relocation of such industries to the periphery of Delhi, despite the loss of livelihoods such a move threatened. Soon after, via another judgement in 1998, the Supreme Court ordered the Delhi Government to convert the city's buses, taxis, and autorickshaws to CNG by 2001, which it considered a cleaner fuel based on the advice of the EPCA. These developments signalled the height of judiciary-led urban environmental shifts in Delhi, as activists came to rely heavily on the courts to push the state to regulatory oversight, enforcement, and even more forward-looking policy changes. In 2010, the National Green Tribunal was notified, with five benches tasked with considering environmental issues brought before the courts, while also disposing them expeditiously. Seen holistically, with its increased role and scope of action, the judiciary has brought into "Indian environmental jurisprudence numerous principles of international environmental law. These include the polluter pays principle, the precautionary principle, [and] 'the principle of inter-generational equity'" (Rajamani and Ghosh, 2012, p150).

One element of atmospheric jurisprudence in India—within which Delhi has been a kind of laboratory—has been the turn from statutory cases, that is, where the relevant regulatory authority brings specific violation of environmental laws to court, to claims made under the Constitution's Article 21 that guarantees right to life of citizens. Access to a clean environment has been interpreted by the courts as a fundamental right, and threats to it can thus be brought to the courts not only by those directly affected by pollution *qua* nuisance, which was the case earlier, but by any party in the larger public interest. The preference for rights-based environmental jurisprudence is in part because of the time statute cases typically take, as appeals move from the lower to higher courts. In the words of Shibani Ghosh, an environmental lawyer who works at CPR,

> if you had to activate the Air Act, you will have to go to Magistrate's Court. Now who has the time and resources? Not only that . . . you're not going to get the impact that you want, because the order may come after 17 years.
>
> (Personal communication, 2020)

Ghosh refers here to a case related to criminal proceedings against a polluter that took those many years to resolve. In contrast, with rights-based cases, a petitioner may approach the high court or Supreme Court directly, bypassing the lower courts. Overall, Ghosh believes environmental petitioners have seen "more victories in rights-based cases than [they've] in statutory cases" (Ibid.). Echoing these thoughts, at a public meeting in 2018, Anumita Roychowdhury of CSE credited the judiciary for its environmental orders related to Delhi, and by extension, making the capital the catalyst for national action on air pollution. At an online roundtable in August 2020 that one of us participated in, Roychowdhury praised the judiciary for its recognition of the right to health.

Other observers are less sanguine about the judiciary's role in environmental issues. They argue that courts often pass orders that are practically impossible to implement on part of the executive, which has to balance several different agendas and exigencies *in addition* to the environment. In the residue burning matter, despite NGT orders and serious efforts, states have not been able to stop or even significantly lessen the fires. The issue is highly complex and silver-bullet techno-solutions, like the Happy Seeder, favoured by governments (see Chapter 3) do not account for things like the problems of landless farmers, the increasing salinity of land, and the larger agricultural crisis which has eroded farmer savings. Another concern is related to the peculiar and inconsistent workings of the public interest litigation (PIL), the preferred vehicle of environmentalists to approach the courts. In his careful reading of PILs related to environmental concerns, Bhuwania (2018) shows that by relying on expert committees and the *amicus curiae*, or a "friend of the court," in cases of "citywide scope, a relatively small number of stakeholders actually played much of a part in the deliberations." Bhuwania believes that PILs have become a highly "malleable" tool in the hands of judges to impose their particular worldview on disparate urban concerns. This is the reason why some PILs have been endlessly prolonged and scaled up to bring in issues far removed from their original scope, continuing to be mulled over by courts up to three decades from their original filing. Reading Ghosh and Bhuwania together, it seems to us that while the latter's criticism of the PIL is undoubtedly valid, the reason why well-meaning activists have taken the route must be understood in a historical manner in relation to the larger developmentalist slant of the state, the powerful interests with stake in the continued pollution, and the hitherto weak public response to urban environmental concerns. Having said that, it is perhaps time advocates critically examine the implications of the judiciary's roles, and recalibrate their strategies given the increasing public consciousness and mobilisation around urban environmental concerns. We turn to the latter matter now.

Atmospheric citizenship

The liberal theory of the state emphasises its place as separate from the (civil) society as the arbiter of social disputes, debates, and conflicts. Claims made by civil society are seen to flow from the fact of belonging to a particular political unit, or one's citizenship. The historically shifting dynamics linking power and subjects are termed "citizenship projects" by Rose and Novas (2005). Such projects have involved, among others, defining the political community, legal systems, national language, and social welfare. Over time, liberal societies have extended the scope of substantive rights to education, health, employment, and even information. While the debate on citizenship is not new and extends far beyond the scope of this work, there are two elements—urban and biological—in the recent reframing of citizenship that are important to think about what an atmospheric citizenship may mean.

Urban citizenship has been powerfully debated recently by movements organised under the broader Right to the City umbrella (Harvey, 2012). They argue that, first, since cities, by their very constitution, are relatively heterogeneous, issues of inclusion become important, especially in the context of majoritarian and xenophobic bent to certain urban politics. The ability of individuals and groups to belong while being different assumes importance in the relatively cosmopolitan urban scenario. Second, cities are characterised by what Manuel Castells termed "collective consumption," or the socialisation of basic services and infrastructures, organised by the state through extended networks (Pahl, 1978). Substantive urban citizenship implies meaningful access to these infrastructures. However, given the considerable unevenness across class, race, caste, and ethnic lines, marginalised groups often must organise and demand access. In some instances, they mobilise in defence of clandestinely built critical infrastructures—by squatting or tapping electricity wires—but which are deemed illegal by the state (Bayat, 2000). Third, an aspect of urban citizenship is related to the sharing of and caring for commons (Zimmer and Cornea, 2016). These include public spaces such as streets, parks, and squares; remnant ecological systems like urban forests and wetlands; and the atmosphere. In the recent past, the intense pressure to commodify and privatise commons has been resisted by communities and collectives in India and elsewhere.

Broadly speaking, Indian debates on urban citizenship have shown that it is differentiated (Heller and Mukhopadhyay, 2015), that is, highly unevenly accessed; and negotiated, such that claims cannot be taken for granted but emerge, for many communities, through protracted "stories of dwelling and belonging" (Das, 2010). Further, following Wacquant (1996), it should be noted that substantive urban citizenship is not necessarily equal to wealth. Histories of specific neighbourhoods are equally important. *Where* one dwells matters. In Delhi, minority ghettos, unauthorised colonies, and slum-designated areas, especially those at the city's physical margins, are most vulnerable to abjection, disruptions, uncertainties, and environmental risk due to their fraught links with economic and political power. Basic urban and environmental rights then become an arena for long-drawn-out struggles.

To anthropologist Adriana Petryna (2004), the classical idea of citizenship as handed by birth or naturalisation is inadequate to account for the real protections necessary for survival in conditions of severe economic insecurity and state incapacity. In the case of those exposed to radiation during the Chernobyl disaster, she argues that the tie that binds the Ukrainian state to claims for social protection is not their official citizenship, but the state's recognition of bodily impairment due to radiation. As Petryna adds, "the tighter the connection that could be drawn, the greater the chance of securing economic and social entitlement" (2004, p262). Petryna calls this coupling of the body and claims, "biological citizenship." Taking the idea forward, Rose and Novas (2005) describe the growing importance of such forms of citizenship, arguing that "claims on political and non-political authorities are being made in terms of the vital damage and suffering of individuals or groups and their 'vital' rights as citizens" (p441).[7] The heightened consciousness of biological risk, vulnerability, and suffering is further constitutive of group identities, or what they term "rights biocitizenship" (p442). Collective solidarity and organising around health is driven by information and digital media, while also working through more traditional activist strategies of petitioning and protest.

We find that the notion of "atmospheric citizenship" contains both biological and urban moments: that is, a vital relation with bodily harm *and* a concern with city life, planning, and environment. The potential of harm by toxic air—expressed in the popularity of such ways of knowing pollution as cigarettes smoked or years of life lost—increasingly informs claims to the state for clean air and a healthy city, which, in the specific Indian juridical context, makes use of the recognition of environmental degradation as undermining the right to life itself. According to Moe-Lobeda (2016), there are three elements to atmospheric citizenship: the

> [f]irst is the implication that one is a "citizen" of the atmosphere as well as a citizen of varied bodies politic. A second is the implication of moral duty . . . toward the atmosphere. A third implication pertains to the rights of citizens, suggesting that a citizen of the atmosphere has rights related to the atmosphere and the services it provides.
>
> (p38)

The first is a fact of presence, the second of subjectivity, and the third, more than anything else, is historical. Like urban citizenship, a section of the population may not possess the social and cultural capital to *claim* clean air.

In an important social science analysis of Delhi's air pollution, Véron (2006) considered the turn-of-the-century anti-pollution activism and judiciary-led interventions to make three broad arguments. First, in contrast to water, Véron argued that activism was concerned with public action rather than private solutions. Second, it was largely "reflective of a general middle-class bias" and activists had turned to the courts, "mostly to represent middle-class environmental interests" (p2104). Third, Véron found it natural that subsequent court-ordered interventions "worked against the interests of the urban poor, especially against slum dwellers and squatters" (p2104). These findings were consistent with other analysis of judiciary-enforced actions and their links with the urban reordering that defined millennial Delhi (Sharan, 2014). Options for environmental justice seemed foreclosed.

More recent work on urban environmental politics in Delhi (Follman, 2016; Ghertner, 2020) shows a move away from the middle-class agenda, though within limits. Follman's study questions the environmentalism of the poor versus elite environmentalism binary in the case of civil society organisations working for the restoration of the Yamuna River. In the course of their activism, these groups have come to oppose the degradation caused by industry as well as the beautification agenda in the form of slum removal. Follman (2016) notes that

> the predominantly negative picture drawn of middle-class activism under the label of bourgeois environmentalism is . . . at odds with the

diversity of urban environmental activism and the wide spectrum of approaches and concerns taken up by middle-class dominated ENGOs engaged in urban environmental policy-making,

(p6)

finding that "the debate about bourgeois environmentalism . . . overshadows these recent forms of urban environmental activism advocating for environmental protection and ecological restoration, as well as socio-ecological justice in the Indian context" (p16). In their work on the politics around waste-to-energy plants in Delhi, Demaria and Schindler (2016) show that "while environmental politics in urban India has hitherto been understood as the preserve of a bourgeoisie intent on imposing revanchist order and disciplining the poor, [the w-t-e case] demonstrates that environmental politics can foster unlikely alliances among these groups" (p308). These (fraught) possibilities are also observed by Webb (2013), though he adds that inequalities may be reproduced in the division of labour, engagement with the external world, and spaces where diverse agents meet.

Tracking the fate of two court cases, namely, *Arjun Gopal*, which concerned the banning of firecrackers around Diwali, and *Vardhaman*, which came to be about phasing out of old diesel vehicles, Ghertner (2020) sees the rise of an atmospheric citizenship as demands for conditions of liveability at once shared by all of the city's residents. He sees this current phase of air-related activism in contrast to the earlier elitist visions of a segregated city where "nuisance" was mobilised in law to target subaltern bodies and livelihoods in the service of a "clean, green city." Ghertner, however, cautions that earlier logics continue in some form, most noticeably in the ability of the elite to produce segregated atmospheres via various technological solutions, as well as viewing clean air as a birthright, a la "upper-caste Hindus' entitlement to and constitution through purity" (p154). Indeed, there are unequal infrastructures of breathing as gated and air-conditioned communities exclude the poor, who are pushed beyond the city with their homes cleared away for wider roads, newer expressways, and green spaces. Another of the state's preferred solutions—the construction of ring highways—similarly relocates polluting trucks to the city's peripheries, even as the health of residents in these peri-urban areas is conspicuously absent from most conversations. In addition, not everyone can understand pollution-related vocabulary of standards, indices, or the loss of life expectancy, let alone act on them. The devolution of data and advocacy has certainly expanded expertise beyond the state's technical units and older organisations like CSE. But the depth of technoscientific expertise required to make sense of them, as well as resources needed to act on the warnings, implies that most people are outside the expert-led actions against air pollution.

This unequal distribution of air quality is also noted in scientific studies. Kumar and Foster (2009) conclude that relocation of industries and conversion of public transport to CNG has only served to redistribute pollution to Delhi's periphery. It is thus no surprise that a recent report identifies 18 pollution hotspots along Delhi's periphery (Toxics Link, 2014). Further, exposure to particulate pollution is not homogeneous. A study noted a direct relation between income and lung function (Narain and Krupnick, 2007) due to costs of preventive care and treatment. Another study found that 94 per cent of ragpickers surveyed suffered from respiratory symptoms (Ray et al., 2004). Moreover, school children disproportionately suffer from reduced lung function and respiratory symptoms (Ministry of Environment and Forests, 2008), particularly if they live near commercial or industrial areas (Mathew et al., 2013). Another study showed that working women and housewives have the highest daily exposure to poor air quality (Saksena et al., 2007), due to cooking and commuting by walking or using public transport. These studies clearly demonstrate that air quality regulations imposed on Delhi at the turn of the century were, at the very least, unable to ensure breathable air for most of the city's residents. Combined with cloistered, air-conditioned communities, and a state that has historically exported its pollution elsewhere, air has been socially and economically conditioned so as to occupy distinct spheres of purity and pollution. As Choy (2016) reminds us, "breathing together rarely means breathing the same."

Despite these inequalities, air retains the ability to affect everyone, to be some kind of an equaliser. In an interview with a TV news channel,[8] Sunita Narain of CSE spoke about the increased concentration of ozone during the summer months, adding that, by its very nature, it seeks and spreads to less polluted, presumably wealthier, parts of the city. No matter how wealthy one is, following Narain, it is almost impossible to completely insulate oneself from polluted air, and even this effort has certain side effects, including a vitamin D deficiency from lack of exposure to sunlight alongside. Ahmann (2020) suggests that the atmosphere as the hinge of environmental politics implies a constant tension between "connection and difference" (p465). It is seen at some moments as "everyone's problem," and at others as something that affects people differently, depending on where one is. In our experience too, air is at once a shared and differentiated condition. It is also an umbrella with which multiple other elements are entangled. Claims for cleaner air, for instance, encompass concerns about the future of urban elements like transportation and industry alongside demands for access to healthcare. It is thus no surprise that pollution has moved up from a sporadic matter of concern to one of AAP's top priorities, joining affordable healthcare on the list. In the next section, we consider how advocates understand these possibilities of citizenship and the challenge of inequalities to develop a new praxis around air.

Translations and collaborations

As we have shown earlier, the complex of media, civic environmentalism, and individual protective action creates an "atmospheric citizenship," a form of relating oneself to Delhi's air that allows it be collectively understood, articulated, and reworked. In the 1990s, however, a vocabulary based on cross-cutting concepts of sustainability, environmental rights, and the precautionary principle was mobilised to promote anti-poor interventions; rightly inviting the criticism of elite environmentalism. The slow shifts in advocacy and activism that we have outlined in the previous section allow us to ask a different set of questions in the present: How open is the debate to different positionalities? Are the concerns of varied populations represented? Faced with anxieties of representation, how does environmental advocacy engage with the multiple unarticulated sites of knowledge and victimhood?

According to Natasha Khanna,[9] a sociology graduate who works at a community mobilisation platform in Delhi, language is a matter of concern. Khanna says:

> Most air quality information [is] available in English. How do you expect people to understand anything that way? CPCB [makes] no real effort in disseminating that information. So in Badarpur [a Delhi neighbourhood where the organisation campaigned], we printed fliers and pamphlets in Hindi. We found out that they were having waste collection problems. Nearby also there's a landfill. People were concerned about smoke coming from there all the time.
>
> (Personal communication, 2017)

Such acts of translation and communication have become increasingly critical to inclusive advocacy, and our interlocuters look to situate global discourses of climate change and sustainability alongside local and regional narratives. To further illustrate this point, we delve into our conversations with two important voices in Delhi's environmental advocacy space.

Sundeep Kumar from HCA,[10] an umbrella group with scores of members, has an almost two-decade-long experience of environmental organising, having been part of Greenpeace in India and the Mediterranean region. Sundeep's expertise is in promoting environmental causes by finding the most effective entry points to produce maximum public impact. He believes that environmentalists in India encounter strong institutional apathy and inertia, and advocates must therefore locate the most precise pressure points to push the system to action. How activists articulate their concerns is important and language is critical. Sundeep described to us the three strategies that seemed to work best. First, new idioms of representing the dangers of pollution were important to think through. He felt that the scientific category of mortality was not an effective invitation for people to care about air pollution:

> [People] are like, *sabko marna hai ek din* [everyone will die one day]. But if we start addressing the lifestyle angle, or the quality of life, like it's not about people dying, but what kind of life are you living? What kind of air are you breathing? That, I think people can get.
>
> (Personal communication, 2019)

Second, HCA looks for political openings when the gaze of the world, so to speak, is on India. During the Under-17 football World Cup in India in 2017, for instance, the group "bought about 50 tickets . . . and went with placards, unfurled a banner, saying our goal is clean air, and then got the conversation going around that. (Ibid.)" They also saw the Delhi Marathon as an opportunity to engage the public in a conversation on the deleterious impacts of physical activity during peak-pollution season. Sundeep finds the activism on Diwali firecrackers similarly productive:

> From a data point of view, one could say that it's very small . . . the major contributors are vehicular pollution or crop burning . . . but people talking about [firecrackers] does contribute . . . in the previous years there was no conversation around air pollution. We made that happen.
>
> (Ibid.)

Third, the alliance has been working hard to translate scientific studies into vernacular languages to bring regional media into the conversation via outreach efforts outside the metros. In particular, their messaging is increasingly targeted at younger residents of smaller towns with the use of social media and contextual vocabularies. Sundeep considers collaborative practice as a critical element of heightened awareness and action around air pollution, adding that it is important for allies to drop their individual brand from collective movements. He sees his role as "convening and coordinating," bringing together environmentalists, lawyers, data scientists, citizen scientists, and community mobilisers. Sundeep is upbeat: "Everybody brings in their resources, their networks, and their skills and we make things happen with whatever we have" (personal communication, 2019).

The second voice we present is of the late Ashutosh Dikshit, erstwhile CEO of URJA, an alliance of over 2000 elected RWAs from across Delhi (see Chapter 2). RWAs have been the chosen institution representing the interests of homeowners and residents of the city's neighbourhoods for a while, but during the Sheila Dikshit government (1998–2013), they were made partners in urban governance through the delegation of functions from state departments. While there remain questions about the effectiveness and progressiveness of these shifts, RWAs have certainly become an important stakeholder in governance and in the city's politics. In this vein, URJA and Ashutosh were part of a citywide movement against high electricity rates charged by private distribution companies. The wider protests on this issue, combined with Sheila Dixit's steadfast defence of the companies, hastened the end of her 15-year rule in Delhi, while bringing a new opposition force in public view that later morphed into the AAP.

It was around this time that URJA threw its weight behind air pollution advocacy, bringing the conversation to RWA spaces, and drawing on their collective weight to push the state to act. While Ashutosh did not consider URJA an activist organisation, since they worked with both the government and civil society, he credited activists for "making noise," which brought issues to people's notice (personal communication, 2017). Through the Delhi Clean Air Forum in 2016, URJA brought together RWAs, "religious institutions, expert committees, heads of schools to sit together and discuss [the issue]" (Ibid.). Like Natasha and Sundeep, Ashutosh too realised the importance of translation since the meaning of pollution varied depending on a person or group's specific experiences. While garbage and plastic burning were important to those in one part of the city, construction dust was the problem for others. The key to Ashutosh was to tie these particularities together into a larger narrative. He said:

> In Mundka people find sewer water overflowing, entering their homes and seeping into food. *Log dekhte hain, log samajhte hain* [people see and understand the issue], these relatable, local-level issues are brought to notice.
>
> (Ibid.)

Ashutosh was engaged in the messy work of what he called "last-mile governance," and he distanced himself from some of the anti-poor bias that has characterised middle-class RWAs in the past (Ghertner, 2012). On garbage burning, for instance, he argued against blaming the poor:

> Most RWAs dislike when security guards burn plastic during winter . . . but give them minimum wages, at least! Give them warm jackets,

gloves and a proper uniform for winter. If you don't provide all this, how will a man keep himself warm?

(Personal communication, 2017)

As we've discussed in Chapter 2, URJA also conducts public surveys and air pollution monitoring and shares its data with the wider public, so that a unified environmental lobby can be created.

Conclusion

The narratives and arguments presented in the chapter suggest that there are significant shifts in air pollution advocacy in the last 30-odd years. While the judiciary remains an important agent of atmospheric governance, the larger emphasis has been decidedly on broad-based campaigning and mobilisation to push governments at various scales to acknowledge the problem, as well as to strengthen institutions to effectively implement regulations. Successful movements demand an openness to work with differently situated agents and willingness to translate across silos. A crucial focus of environmental advocacy, therefore, has been on creating larger alliances to push environmental causes, while also listening to different voices, and pursuing inclusion across space and communities. The latter effort is far from straightforward, given the uneven capacity of different individuals and groups to participate in the debate. This unevenness is related to understanding and articulating scientific data and technical concepts on the one hand, and possessing the resources to participate in long-drawn struggles given the exigencies of everyday life in the city. It should not be altogether surprising to find only those with the economic and cultural capital to be a part of the year-long debate for years on end.

Given the very real constraints, we consider it unfair on our interlocutors and others who have devoted their energies to campaign for cleaner air to be labelled outright as elite environmentalists. We believe it is important to consider their ideas of change, strategies, and tactics. In this scenario, a different set of questions must be asked: How open are they to different positionalities? And are the concerns of varied populations represented? It is here that we see reflexivity and collaborations with the potential to change the terms on which environmental politics proceeds in Delhi and beyond. We see the debate on what constitutes the problem being expanded to fold in an appreciation of the different elements of the urban economy, diverse ways of being, and connections between the city and the wider region. Atmospheric citizenship, in short, has the potential to inform and push environmental politics in more progressive directions with its recognition of air's shared but differentiated condition, its varied experiences and situated knowledges, and the political stakes of institutional architectures and shifts therein.

Notes

1 Rohit Negi, fieldnote, 2 November 2019.
2 UNEP document. Available at www.shorturl.at/cpDIM (Accessed 21 April 2020).
3 Sayani Gupta (pseudonym), researcher with a public health NGO, personal communication, 2019.
4 Sayani Gupta, personal communication, 2019.
5 Literally, the "party of the common man"
6 From Gogoi's intervention at a panel on air pollution at CPR Dialogues, Delhi, 4 March 2020.
7 Vital here is used both in the biological sense as relating to life, and as something necessary.
8 Primetime with Ravish Kumar, NDTV India, 22 April 2019.
9 Pseudonym.
10 The names of the person and organisation have been anonymised.

References

Ahmann, C. (2020) 'Atmospheric coalitions: shifting the middle in late industrial Baltimore', *Engaging Science, Technology, and Society*, 6, p462–485.

Armitage, D., De Loë, R. and Plummer, R. (2012) 'Environmental governance and its implications for conservation practice', *Conservation Letters*, 5(4), p245–255.

Bayat, A. (2000) 'From "dangerous classes" to "quiet rebels": politics of the urban subaltern in the global South', *International Sociology*, 15(3), p533–557.

Bhuwania, A. (2018) 'The case that felled a city: examining the politics of Indian public interest litigation through one case', *South Asia Multidisciplinary Academic Journal*, 17. Available at https://journals.openedition.org/samaj/4469 (Accessed 3 April 2020).

Bolia, N. and Khare, M. (2018) 'A passage to clean air', *Indian Express*, November 9.

Choy, T. (2016) 'Distribution', *Society for Cultural Anthropology*. Available at https://culanth.org/fieldsights/distribution (Accessed 24 September 2020).

Das, V. (2010) *Citizenship as claim; or, stories of dwelling and belonging among the urban poor*. Dr. B. R. Ambedkar Memorial Lecture, Ambedkar University, Delhi, 23 August.

Demaria, F. and Schindler, S. (2016) 'Contesting urban metabolism: struggles over waste-to-energy in Delhi, India', *Antipode*, 48(2), p293–313.

Du, X. et al. (2020) 'Cross-state air pollution transport calls for more centralization in India's environmental federalism', *Atmospheric Pollution Research*, 11(10), p1797–1804.

Ferguson, J. (2006) *Global shadows: Africa in the neoliberal world order*. Durham, NC: Duke University Press.

Follman, A. (2016) 'The role of environmental activists in governing riverscapes: the case of the Yamuna in Delhi, India', *South Asia Multidisciplinary Academic Journal*, 14. Available at https://journals.openedition.org/samaj/4184 (Accessed 28 June 2020).

Ghertner, D.A. (2012) 'Nuisance talk and the propriety of property: middle class discourses of a slum-free Delhi', *Antipode*, 44(4), p1161–1187.

Ghertner, D. A. (2020) 'Airpocalypse: distributions of life amidst Delhi's polluted airs', *Public Culture*, 32(1), p133–162.

Ghosh, S. (2015) *Reforming the liability regime for air pollution in India*. New Delhi: Centre for Policy Research. Available at http://re.indiaenvironmentportal.org.in/files/file/Reforming%20the%20Liability%20Regime%20for%20Air%20Pollution%20in%20India.pdf (Accessed 19 July 2020).

Harvey, D. (2012) *Rebel cities: from the right to the city to the urban revolution*. New York: Verso.

Heller, P. and Mukhopadhyay, P. (2015) 'State-produced inequality in an Indian city', *Seminar*, 672, p51–55.

Indian Express (2019) 'Captain Amarinder Singh terms Kejriwal's claims on stubble burning "shameless lies"', *Indian Express*. Available at https://indianexpress.com/article/india/capt-terms-kejriwals-claims-on-stubble-burning-shameless-lies-6095792/ (Accessed 23 May 2020).

Indian Institute of Management Lucknow (2010) *Evaluation of central pollution control board*. Available at www.indiaenvironmentportal.org.in/files/Rpt_CPCB.pdf (Accessed 4 April 2019).

Karnad, R. (2020) 'The coronavirus offers a radical new vision for India's cities', *New Yorker*, 13 April. Available at www.newyorker.com/news/dispatch/the-coronavirus-offers-a-radical-new-vision-for-indias-cities-pollution (Accessed 21 August 2020).

Khanna, I. and Sharma, S. (2020) *Could the national capital region serve as a control region for effective air quality management in Delhi?* New Delhi: Collaborative Clean Air Policy Centre.

Kumar, A. (2017) 'Law aiding Monsanto is reason for Delhi's annual smoke season', *Sunday Guardian*, 30 December. Available at www.sundayguardianlive.com/news/12191-law-aiding-monsanto-reason-delhi-s-annual-smoke-season (Accessed 5 April 2020).

Kumar, N. and Foster, A. D. (2009) 'Air quality interventions and spatial dynamics of air pollution in Delhi and its surroundings', *International Journal of Environment and Waste Management*, 4(1–2), p85–111.

Lidskog, R. and Sundqvist, G. (2002) 'The role of science in environmental regimes: the case of LRTAP', *European Journal of International Relations*, 8(1), p77–101.

Marrapu, P. et al. (2014) 'Air quality in Delhi during the commonwealth games', *Atmospheric Chemistry and Physics*, 14, p10619–10630.

Mathew, J. et al. (2013) 'Environmental and occupational respiratory diseases-1057 correlation between air pollution and respiratory health of school children in Delhi', *World Allergy Organisation Journal*, 6(1), p55.

Ministry of Environment and Forests (2008) *Study on ambient air quality, respiratory symptoms and lung function of children in Delhi*. Available at http://cpcb.nic.in/upload/NewItems/NewItem_123_children.pdf (Accessed 29 October 2017).

Ministry of Environment, Forests and Climate Change (2019) *National clean air programme*. Available at http://moef.gov.in/wp-content/uploads/2019/05/NCAP_ Report.pdf (Accessed 2 August 2020).

Moe-Lobeda, C. (2016) 'Climate change as climate debt: forging a just future', *Journal of the Society of Christian Ethics*, 36(1), p27–49.

Narain, U. and Krupnick, A. (2007) *The impact of Delhi's CNG program on air quality*. Discussion Paper, RFF DP 07-06. Washington, DC: Resources for the Future.

Negi, R. (2019) 'Why Indian anti-pollution activists must take politics seriously', *Scroll.in*, 23 November. Available at https://scroll.in/article/943957/ why-indian-anti-pollution-activists-must-take-politics-seriously (Accessed 21 September 2020).

Negi, R. and Srigyan, P. (forthcoming) 'Peopling technoscience: the sciences and publics of air pollution in Delhi', *Dialogue: Science, Scientists and Society*.

Pahl, R. E. (1978) 'Castells and collective consumption', *Sociology*, 12(2), p309–315.

Petryna, A. (2004) 'Biological citizenship: the science and politics of chernobyl-exposed populations', *Osiris*, 19(4), p250–265.

Rajamani, L. and Ghosh, S. (2012) 'India', in Lord, R., Goldberg, S., Rajamani, L. and Brunnée, J. (eds.) *Climate change liability: transnational law and practice*. Cambridge: Cambridge University Press, p139–177.

Ray, M. R. et al. (2004) 'Respiratory and health impairments of ragpickers in India: a study in Delhi', *International Archives of Occupational and Environmental Health*, 77(8), p595–598.

Rose, N. and Novas, C. (2005) 'Biological citizenship', in Ong, A. and Collier, S. J. (eds.) *Global assemblages: technology, politics, and ethics as anthropological problems*. Malden: Blackwell, p439–463.

Saksena, S. et al. (2007) 'Daily exposure to air pollutants in indoor, outdoor, and in-vehicle micro-environments: a pilot study in Delhi', *Indoor and Built Environment*, 16(1), p39–46.

Sharan, A. (2014) *In the city, out of place: nuisance, pollution and dwelling in Delhi c. 1850–2000*. New Delhi: Oxford University Press.

Singh, S. (2018) *The great smog of India*. New Delhi: Penguin.

Soroos, M. S. (1991) 'The atmosphere as an international common property resource', in Nagel, S. S. (ed.) *Global policy studies: international interaction toward improving public policy*. London: Springer, p188–220.

Swyngedouw, E. (2005) 'Governance innovation and the citizen: the Janus face of governance-beyond-the-state', *Urban Studies*, 42(11), p1991–2006.

Toxics Links (2014) *On the edge: potential hotspots in Delhi*. Available at http:// toxicslink.org/docs/Report-On-the-Edge.pdf (Accessed 24 February 2018).

Véron, R. (2006) 'Remaking urban environments: the political ecology of air pollution in Delhi', *Environment and Planning A*, 38(11), p2093–2109.

Wacquant, L. J. (1996) 'The rise of advanced marginality: notes on its nature and implications', *Acta Sociologica*, 39(2), p121–139.

Webb, M. (2013) 'Meeting at the edges: spaces, places and grassroots governance activism in Delhi', *South Asia Multidisciplinary Academic Journal*, 8. Available at https://journals.openedition.org/samaj/3677 (Accessed 4 August 2020).

Zimmer, A. and Cornea, N. (2016) 'Introduction: environmental politics in urban India', *South Asia Multidisciplinary Academic Journal*, 14. Available at https://journals.openedition.org/samaj/4247 (Accessed 12 July 2020).

5   Smart business

Ecopreneurship and its dilemmas

Air is where water was 25 years ago in Delhi . . . middle- and upper-income families boiled water to remove dangerous toxins. Today, these same families have fitted their apartments with elaborate reverse-osmosis filters. With time, air filters will get there too.[1]

As we have described in the previous chapters, the concern over Delhi's air has peaked since 2014 and reaches crescendo during the winters when smog envelops the region for several weeks. All of the region's residents viscerally experience the smog, and many of them are forced to seek out healthcare due to asthma, wheezing, and other respiratory ailments. The enhanced visibility of pollution, its health impacts, and the larger alarm around toxic air that plays out via the press and social media produces demand for ready fixes for individuals and families. While people have adapted by closing windows and fashioning face coverings for the longest time, today several—far more effective—technological solutions are at hand to paying customers. Never before has the possibility existed to precisely map the properties of one's proximate air through low-cost monitors or to install machines that suck up particulate matter and spew out clean air.[2]

It is estimated that the market for air purifiers alone is worth over Rupees 450 crore (USD 60 million) in India.[3] Several multinationals and smaller local startups market various kinds of purifiers, ranging from a few thousand to over a hundred thousand rupees, depending on functions and brand. In addition, car air purifiers and paints that promise to remove pollutants from household air are available to the discerning customer who wants clean air at all times. Users of many personal air purifiers can also "know" their proximate air as data, since the sensors on the purifier transfer details of indoor air quality real time to a phone via its app (Figure 5.1) In short, a huge and growing market in various technologies and gadgets has emerged in the last half-decade in India, and given its air quality, the disproportionate

Figure 5.1 The indoor atmosphere is shown as sharply insulated from the outdoors by an air purifier

Source: Rohit Negi.

media coverage, and the purchasing power of many residents, Delhi contributes strongly to this market. This chapter focuses on these developments to understand the nuances of the scramble to provide innovative entrepreneurial solutions to air pollution, outlining both the larger processes that explain its emergence as well as the specific stories of some of those engaged in the business.

To be sure, to highlight the tendency of capitalism to market remedies to problems that it is responsible for is a productive entry point into this situation. Relatedly, the neoliberal impulse to seek individualised solutions to social and political concerns, such as pollution, is yet another forceful critique. However, in keeping with this book's emphasis on building concepts *from* ethnographic work rather than pregiven frames, we argue that the commodification of air is a complex process that far exceeds the narrow framings within which it is otherwise. We situate these developments in Delhi within a different storyline, one at the intersection of specific processes in the Indian urban context: first, the ongoing scientisation of urban knowledges linked to the imperative to make cities "smart;" second, the emergence of what Irani (2019) calls "entrepreneurial citizenship," or the shaping of a specific subjectively tied to capitalising technical solutions to social concerns; and third, the growth of urban environmentalism that leads individuals to make ecological concerns vital to their life projects.

In the last decade, these elements have come together in the context of Delhi's worsening air quality to produce what is known in the management literature as "ecopreneurship," or the simultaneous pursuit of environmental good and profits. Tracing the trajectories of our interlocuters who are engaged in ecopreneurship in Delhi, we show that these are highly contextual paths at the intersection of personal anxiety, activism, and profit. Delhi's ecopreneurs are motivated by personal interests but aim to connect with wider publics and contribute beyond the narrow profit-motive. The chapter further argues for these moves to be taken seriously to understand why the boundary is increasingly blurred between otherwise contradictory impulses: selling a commodity on the one hand, and devoting energies to public welfare on the other.

Ecopreneurship and its critiques

The intellectual history of entrepreneurialism, though old and varied, coalesces around a few basic ideas, contributed chiefly by Joseph Schumpeter (Śledzik, 2013), related to the elements of creativity, innovation, and risk-taking in the pursuit of generating profits. Ecopreneurship is a more recent term that combines ecology and entrepreneurialism. It seeks to succinctly capture the essence of the creation of a new space within business practice over the last 40-odd years, as entrepreneurs respond to the environmental movement's foregrounding of concerns ranging from biodiversity loss to global warming and water and air pollution. According to Schaper (2002), the idea emerged in business scholarship in the 1970s, around the time an influential paper (by J.B. Quinn) highlighted an approach to environmentalism that went against the grain at the time. Quinn argued that the increasing societal concern with environmental problems needed to be seen as an opportunity—to create new products and expand markets—rather than an economic drain. Over time, as the environmental agenda grew, and regulations and public oversight tightened, corporations have responded through various means. Kim Fortun (2009) shows the absence of substantive shifts on the part of the chemical industry, for instance, which has resisted change through various means. Others have pointed out attempts at "greenwashing," or token changes to business-as-usual while playing them up through slick marketing. Oil companies are particularly adept at these strategies, given their investment in destructive extractive activities and the substantial advertising budgets at their disposal (Cherry and Sneirson, 2012). At the same time, other businesses have viewed this as an opportunity to retool and present themselves as "green" alternatives to environmentally mindful consumers. Sometimes the shifts are connected to the practice of what is termed as "intrapreneurship," or the incentivising of innovation within existing businesses. But very often, and this is the focus of this chapter, ecopreneurship involves individuals or small collectives developing new business and products in the environmental domain. This process has grown manifold since the 1990s, as regulatory shifts from the top and consumer demand from below make green consumption attractive to corporations and individual entrepreneurs. In short, ecopreneurship may be viewed as the process of profit-oriented innovation towards environmentally favourable ends.

In the extant business literature, ecopreneurship is largely seen in positive terms as a win-win solution. This scholarship views environmental problems as a result of market failure, and, in essence, places faith in the market itself to provide solutions. Thompson et al. (2011) define ecopreneurship as a "focus on resolving environmental degradation through the creation of new products, services and markets" (p222). The ecopreneur is considered critical to the process since these are relatively new ideas and require innovators open to experimentation. One difference between entrepreneurs and ecopreneurs, highlighted by Farinelli et al. (2011), is that to the latter, profits are not the end-all. The authors argue that ecopreneurs are "intrinsically motivated. . . [and] consciously aim at ensuring a more sustainable future" (p43). According to Mary Phillips (2012), ecopreneurs combine "the drive, ambition, creativity and risk-taking of the conventional entrepreneur with a concern for the environment" (p798). O'Neill and Gibbs (2016), however, point to the reliance in this literature on a discourse of the "entrepreneurial hero" through the use of tropes like the "maverick" or "opportunist" individual. They call attention to the wider contexts within which ecopreneurs operate, especially the "messy realities" of having to care for the environment while working with the capitalist imperatives of growth and profit. They find that ecopreneurs are more likely to take a pragmatic position on tricky matters as opposed to a puritannical environmental one. At the same time, there were cases of more radical forms of ecopreneurship, where individuals "combined with other activities, including campaigning, food growing, caring, family life and other projects . . . [to] act as educators or change agents" (p15).

On the other hand, critical social science literature on ecopreneurship is less sanguine. Its critique of ecopreneurship can be broadly grouped into three tendencies. The first draws on Marxist political economy—and in particular, the idea of the "metabolic rift" (Foster, 2000)—to argue that capitalism leads to environmental degradation as a rule. A similar argument has been recently made by Jason Moore (2015), who shows how human history is entwined with nature, and capitalism represents a specific world historical–ecological moment in this process, one where nature is appropriated and exploited. To then expect it to provide lasting ecological solutions is a fallacy. The second, and related, critique considers the more recent ecopreneurial developments as signifying what Jesse Goldstein (2018) terms the "new green spirit of capitalism," where value is being increasingly extracted from the productive energies of environmentalism while leaving the system intact. In other words, the progressive push towards ecological thinking is channelled by capital to further engineer profits, leading to the conflation of environmentalism and consumerism, eliding calls for systemic change (Luke, 1998). Third, it has been argued that some of the new technologies of amelioration available to individuals are geared towards self-care (Kimura, 2016), precluding meaningful engagement with the political process, which is seen as more likely to bring lasting change.

There is then a wide rift between the business and critical literatures on ecopreneurship. Our purpose in this chapter is not to suture the chasm. Rather, we view Delhi's ecopreneurs as embodying and trying to muddle through these contradictions as reflexive agents. Further, most of the literature cited earlier is built on research in the global North, and in our work we find several dilemmas that are quite specific to our historical-geographical context. In this chapter, we work with these particularities to *situate* Delhi's airpreneurs, while contributing a study from the Global South to the larger literature on the theme. We ask the following questions: What explains the emergence of air pollution-related ecopreneurship in Delhi?

What are Delhi's ecopreneurial "types" in terms of individual trajectories and arrivals? How are the contradictory concerns of environmentalism and profit-making negotiated? And how do we think public interest with the ecopreneurial turn in environmentalism?

We show that ecopreneurship in Delhi is positioned at the intersection of highly specific processes that have shaped not only its practice but also subjectivities of those involved. Thereafter, the chapter delves into the stories of ecopreneurs to identify common patterns and also the exceptions to how they look to run successful businesses while also working to reduce air pollution. It describes the narrative and performative strategies these individuals deploy as they negotiate the tension between profit and environmentalism. Finally, we show how, motivated by public interest, but operating within constraints, individuals and collectives reframe the meaning of ecopreneurship as they look to build participatory praxis. Since the chapter looks specifically at these developments around air pollution, in what follows, we use the terms "airpreneurship" and the "airpreneur" to discuss the case of Delhi.

Situating airpreneurship

Broadly speaking, there are two trajectories of airpreneurship in Delhi. One involves individuals with business background and training moving to air pollution as a result of chance bodily encounters, while the other is about technical professionals who have arrived at the entrepreneurial space as a result of the confluence of highly specific contemporary processes. We show, in this section, how a concern with the environment, which is common to both groups, intersects with larger urban processes to produce these airpreneur types, with their particular character and possibilities for public interest and environmental justice.

Over the last three decades, environmental concerns, debates, and interventions have come to be central to what it means to be urban in India. This is especially true of Delhi, where scholars have shown the varied contestations around water—both in the form of the river and drinking water (Follman, 2016)—urban green spaces (Sivaramakrishnan, 2017), waste (Gill, 2009; Demaria and Schindler, 2016), and climate change (Hughes, 2013). Through these studies, some larger patterns in the environmental politics of the Delhi region emerge, which include the critical role of the judiciary, the largely middle class-led nature of activism, peripheralisation of materials and activities viewed as nuisance, and an increasing reliance on techno-solutions (de Bercegol and Gowda, 2019). Regardless of the specific emphases, the space of urban environmentalism has grown tremendously, and attracts several individuals who contribute their time, resources, and expertise to one or the other issue. Doing so at this specific moment in history, with the ever-growing importance of digital mediums, involves working closely with data in their various manifestations. Seeing or breathing toxic air isn't enough to make a point; pollution has to be captured as data and calibrated against extant standards to be recognised.

The smart urbanism agenda that explains these shifts has been constitutive of new kinds of urban policy, planning, and market initiatives. Cities around the world have retooled governance such that data-drivenness has become the hinge for the operation of various aspects of urban life, from transportation and design, to food and surveillance. One emphasis has been on using sensing technologies to generate various kind of data and link these to form "big data" along the lines of the "internet-of-things." In our context, the Government of India's Smart City Mission aims to develop greenfield cities and retrofit existing ones according to the principles of data-drivenness, networking, and speed, while private companies develop platform solutions from transportation (Uber, Ola, etc.) to services (Urban Clap, etc.), food delivery (Zomato, Swiggy, etc.), and medical care (Practo, etc.). According to Ayona Datta (2018), the smart urban agenda requires two moves: first, to bring a large population into the digital domain, and second, showing them "how to perform as 'smart citizens' in order to contribute to the 'success' of the smart city" (p406). Scholars remind us that there is not one kind of smart urbanism but several context-mediated ones. While critical research points out several problematic aspects of these developments—ranging from unequal access to corporate domination and state control—Coutard and Guy (2007) argue that "adaptation, contestation and even resistance" (p730) are as much a part of the larger picture.

Sadowski (2020) sees smart urbanism as "managerial methods for entrepreneurial ends delivered via smart solutions" (p4). David Harvey had shown a while back (1989) how urban governments were becoming entrepreneurial in the neoliberal era, competing with each other to attract and retain private investment. With smart urbanism, we now find a move from those entrepreneurial cities to "cities *for* entrepreneurialism" (Levenda and Tretter, 2019), that is, cities become the arena for private entrepreneurial activities, and facilitating these activities becomes a core element of urban policy. These processes have extended to the environmental space as the state looks to facilitate smart-entrepreneurial solutions to environmental problems.

Entrepreneurialism encouraged by smart urbanism works through specific sites and mechanisms, and produces subjectivities consistent with belief in market and technological solutions. It is supported by corporations, venture capital (VC), and governmental subsidies that budding entrepreneurs look to tap into. A part of this ecosystem is "social entre-preneurship," comprising enterprises that aim "to fulfil a social purpose as well as achieving financial sustainability through trading" (Haugh, 2005, p3). As Lily Irani (2019) notes, development-oriented entrepreneurs are tired of the bureaucratic and consultative ethics of Indian policy and practice, and are attracted by the notions of speed and vision promised by the millennial smart-entrepreneurial culture, where "vision is the distant promissory horizon to set for oneself, whereas speed is the means by which to narrow that distance as energetically as possible" (Sundar Rajan, 2006, p88). One of the sites at which horizontal—among entrepreneurs and technical experts—and vertical—with VCs and government departments—interactions take place is the hackathon, or "an intense, multi day event devoted to rapid software production" (Irani, 2015, p803). Individual participants at these events are most often connected to the worlds of software development and design; they take time out from their usual routine, work collaboratively, ideate, and create a prototype—the "demo"—which may then be picked up by private capital or the state. Hackathons underscore "the belief that urban issues are solvable through technological fixes, with hackathons leveraging the innovation capacity of a crowd of talented, technically literate citizens to practice. . . 'solutionism'" (Perng et al., 2018, p189).

Beyond the matter of success or failure of these experimental initiatives—including FabLabs, do-it-yourself technologies—the important point is that they produce specific subjects; individuals with an interest in social change, belief in the transformative potential of technology, and faith in the market. Moreover, as Mertia's (2017) reading of a group of open data entrepreneur/activists, who engage with the technologies towards democratic ends, suggests, these individuals "are not just participating in a democratic process in a fixed sense of the term. Rather, they are creating new, technologically mediated and/or affected meanings of democratic citizenship" (p12). It is an empirical question as to the subjectivities, possibilities, and limits of the intersection of environmentalism with smart urbanism and entrepreneurial citizenship. In the following sections, we describe the trajectories of Delhi's airpreneurs, their motivations and narrative strategies, and their positions with respect to questions of environmental justice.

Delhi's airpreneurs

As already mentioned, there has been a scramble to offer technical solutions to Delhi's toxic air in the past half-decade. Primarily, the market for personal masks, air purifiers, and low-cost air quality monitors has enlarged significantly. Increasingly, offices, schools, and malls have installed heavy-duty purifiers for workers, students, and visitors. The PVR chain of cinemas, for instance, advertises the clean air it offers, noting that its "audit*air*iums" add 15 minutes to a film viewer's life. There is also an entire building complex, the Paharpur Business Centre, which claims to provide Alpine air to office-goers through the use of purification technologies and hundreds of pollution-reducing plants. Its CEO, Kamal Meattle, has been featured in multiple media stories and is one of the airpreneurs we feature in this section. Meattle is also recognised for a popular TED-Talk ("How to Grow Fresh Air") on the benefits of three plants—the areca palm, mother-in-law's tongue, and money plant to help reduce indoor pollution. Another of Delhi's airpreneurs is Barun Aggarwal, with whose quote we opened this chapter. Aggarwal is the CEO of Breath Easy Labs, engaged in monitoring and purification of indoor air.

Meattle and Aggarwal, together with two other airpreneurs—Jai Dhar Gupta (Nirvana Being) and Rohit Bansal (Purelogic Labs)—may be categorised, like a few of them self-identify, as "serial entrepreneurs," that is, individuals who seek out potentially new markets in cutting-edge products, and have arrived at the air quality space in the recent past. Nirvana Being is the exclusive importer and retailer of high-quality anti-pollution Vogmasks in India, in addition to heavy-duty medical-grade air purifiers, which they have installed across many places, including in partnership with PVR Cinemas, as described earlier. Purelogic is a more recent entrant in the market for handheld air quality monitors and low-cost household air purifiers. What connects the four airpreneurs are their personal trajectories: they are from business backgrounds and have arrived at airpreneurship after degrees in management or other disciplines abroad. Meattle is an MIT graduate, Gupta has studied at Wharton, Aggarwal at Duke, and Bansal came to India via China after studying in the US. On arrival in India, each of them was hit hard by the polluted air, took it on themselves to provide solutions, and saw the issue as smart business.

The second group of airpreneurs have entered the space via a very different path. They are trained as engineers or data scientists and have arrived through smart urban entrepreneurialism, which we discussed in the previous section. Mrutyunjay Mishra of Indian Open Data Association, who unfortunately passed away in 2019, described how he got into the air pollution space following his participation at the 2015 Kumbhathon organised by MIT in Nashik. The hackathon's specific provocation was to think through the diverse urban challenges posed by rapid population growth, since Nashik represents a productive case study with its manifold population increase during the periodic Kumbh Mela event. It is in the course of participation at the hackathon that networking, ideation, prototyping, and blueprints for continuing discussions emerged. Mishra added,

> I went to Kumbhathon with the idea that can we create a simple, scalable, easy-to-deploy solution that would give real-time environment status of an area for thousands of locations over the next five years? I just collected a bunch of sensors from all over. I bought $2000 worth of sensors. And I bought IoT boards, Internet of Things boards . . . All of those kids . . . oldest one was probably 23, youngest one was 13–14. Now, they wanted to work on this project. They came in, they started playing.
>
> (Personal communication, 2017)

Thereafter, Mishra continued to work with a small group of engineers and invested in their startup that had been incubated at a FabLab at CEPT, Ahmedabad, as they worked towards developing a low-cost air quality monitoring device. By 2017, the startup was already making a decent amount of money from its prototype, and had secured funding from the Gujarat Incubation Centre and from French Tech Ticket to set up an office in Paris. In an interesting turn, which completes the circle, the company recently posted an update on social media pitching their monitor to city governments at a smart city event.

The low-cost monitor sector has attracted a number of the tech/data airpreneurs. Our interlocutors mentioned three ways in which the monitors contribute to the collective understanding of air pollution. First, they help to know one's proximate air. As Nita Soans, CEO of Kaiterra India and a trained architect turned design-practitioner, described, monitors help "people see the air. It [is] not solving the problem. It [is] just a guide, a thermometer or something, or like a weighing scale if you're trying to get on a weight loss programme" (personal communication, 2019). Knowing exactly the air one breathes in turn is supposed to lead consumers to keep themselves and their families safe by taking action—installing purifiers, shutting bedroom windows, or not letting children play outside in the open during bad air days.

The second contribution of low-cost monitors is to provide data to back citizen science and collective actions. According to computer science graduate and founder of Respirer Living Sciences, which has developed the Atmos monitor, Ronak Sutaria, the device

> is a tool by which citizens are engaging with . . . decision-makers. Instead of just going and shouting at the lawmaker saying air quality is

bad, you tell them that the air quality has got worse by 20 per cent, or by 40 per cent month by month.

(Personal communication, 2018)

This is the transition from lay notions of suffering to generalised and data-backed association of harm with source that is widely called "popular epidemiology", a language of redressal without which claims are unrecognised by the state. The third use of low-cost monitors is to assist in a granular mapping of air quality across geographies by filling-in the gaps between the relatively sparse government reference monitors. In this manner, tech-airpreneurs are helping develop a dense picture of pollution with ever more accurate and reliable monitors. This is contrasted to the largely import and sale model of the serial airpreneurs.

Profits versus the environment: narrative and performative strategies

In her interview-based study of ecopreneurs, Mary Phillips (2012) examines the popular notion that business and the environment are opposed to one another: the former implies being supple, even with one's moral principles, in the pursuit of profit (p798), while the latter is seen to be in a state of degradation precisely because of capitalism. To create and run eco-businesses then seems—at first glance—to be paradoxical. Phillips writes, "Contradictions are . . . evident in the ways in which [ecopreneurs] position themselves with and against others, and as simultaneously part but not part of business and green communities" (p795–796). Similar tensions are observed in India with an added element: the legacy of the Gandhian imperative to renounce materialism to do social or environmental good, and many Indian social entrepreneurs cite Gandhi as an inspiration (Acharya, 2019). In response, ecopreneurs deploy some interesting narrative strategies. Many distance themselves from both the purely capitalist and environmental-purist positions, highlighting their pragmatic approach towards the environment, rather than what is seen as the "self-indulgence" of green activists. These narratives are in essence about figuring out "what it means to be neither part of the business nor the green communities" (p810). O'Neill and Gibbs (2016) find a similar tension in their study of ecopreneurs in England and Wales, where ecopreneurs are likely to align with mainstream business ideas, steering clear of activism-oriented environmentalism, which is often disparaged with the use of tropes like "tree-huggers." They realise the constraints within which businesses operate, and, over time, tend to think of themselves as helping make things a "little bit better" than saving the world per se (p1740).

In this section, we discuss the three broad narrative and performative strategies Delhi's airpreneurs employ as they negotiate these tensions. The first is about the airpreneurs' assertion of authenticity of their involvement in the air pollution space by drawing on their own experiences with bad air. Each of the four serial airpreneurs and one of the four tech airpreneurs whom we have tracked have been vocal about air pollution as something very personal and intimate to them. Namita Gupta, the co-founder of Air-Veda, is quoted on the company's webpage:

Asthma runs in my family. My dad, myself and my daughter are all asthmatic. After being in the US for 13 years, we moved to India in Dec 2013 . . . In September 2014 my asthma got much worse, after double courses of oral steroids I realised that air pollution was aggravating my situation significantly. The more I learned about air pollution the more I panicked . . . I bought purifiers and plants. Did they help? Were my children safer now? I felt helpless. I had a choice. Move back to the U.S. or do something about it. I decided to do something about it.[4]

There is a great deal of similarity between her narrative and Jai Dhar Gupta's, who returned to India around the same time. Gupta says,

"I'm an accidental social entrepreneur, trying to save my own life . . . I was training for a marathon in 2013, was at the peak of health, had run 12–14 km and in the evening, was wheezing for the first time ever. The problem kept getting worse, 15–20 days later my lung function was down 40%, I was practically a vegetable . . . it wasn't me but the external environment. I started speaking about it, became an activist, an environmentalist, it was personal."[5]

Similarly, Kamal Meattle told one of us how, in the 1990s, he suffered from reduced lung function, and his doctors strongly advised him to leave Delhi. Unwilling to give up family ties and social networks in the city, he instead sought advice from IIT-Delhi faculty to find technical solutions to pollution, which led him to develop the Paharpur centre, described earlier (personal communication, 2018).

On the Purelogics webpage, there is a series of graphics that depict company director Rohit Bansal's arrival on the scene.[6] Bansal moved to Beijing in 2008 and developed sinusitis. Physicians identified the city's polluted air as the cause of the affliction. He then went on to educate himself on the specifics of air pollution, before settling on monitoring as his area of professional interest. With help from Chinese engineers, he developed a scalable prototype in 2018, which his company now sells, and a network of these monitors also supplies the data for the AQI India app which Purelogics operates. On the other hand, Jeff Smith of Ambient Monitoring, Nita Soans of Kaiterra, and Barun Aggarwal talk about their children's struggles with Delhi's air being a motivating factor for them entering the airpreneurial sector. At a panel in December 2018,where one of us (Rohit) was a co-panellist, Aggarwal repeatedly mentioned that he had to rush from the event to take a family member to Dr Arvind Kumar, a well-known pulmonary specialist and an outspoken anti-pollution activist. Such assertions of a personal stake in clean air are an important narrative element to establish the authenticity and motivation that take their interest beyond just the profit-motive.

We find a second narrative strategy related to one's identity as an entrepreneur and an activist. Many of the airpreneurs, especially of the first type, are heavily involved in advocacy, litigation, and public activism around air pollution. Jai Dhar Gupta is the co-founder of Help Delhi Breathe, Barun Aggarwal of Care for Air, and Namita Gupta has been involved with the #IBreathe campaign. These initiatives have organised online campaigns, educational programmes in schools, and even protests in Delhi's prominent public spaces like Jantar Mantar. At one of these events in January 2016, we heard Gupta speak, in front of a large AQI display, about the change public pressure can make. The event attracted hundreds of concerned Delhiites, and received wide media coverage. In 2018, working alongside other citizen groups, the collective installed a pair of giant artificial lungs outside the Sir Gangaram Hospital before Diwali. The lungs turned black within a week, illustrating the prevailing hazardous air quality in the city. Like before, this installation was covered by almost all major newspapers and television news. In the interview cited earlier, Gupta described himself as an environmentalist and an activist. Meattle too, considers himself first as an environmentalist and then an entrepreneur. He has previously engaged with the campaigns for clean fuel and recyclable packaging, working his considerable social networks towards green regulations. He says: "when I do these things, I am not doing these for myself; I am doing them for society. That's what keeps me going. It makes me energised" (personal communication, 2018). Namita Gupta too sees herself as an activist. Learning from her experience with toxic air, she realised that "data can help create more aware citizens who are willing to change their own behaviour. She became an evangelist attending marches and events, and started using social media to drive awareness about air quality."[7] Being in the business for personal reasons and crossing over into the realm of public activism is then an important element of contemporary ecopreneurial subjectivities in Delhi.

At the same time, and in line with the findings of Phillips (2012) and O'Neill and Gibbs (2016), in the cases of two tech ecopreneurs, we see a distancing from advocacy. In the case of Ronak Sutaria, this move is slow.

Sutaria had some experience of the monitoring technologies from his time in California, but got involved more seriously with them in partnership with the data journalism portal *IndiaSpend*. He worked with the group from 2015 for 2 years, developing, installing, and calibrating low-cost monitors; analysing the data received from them; and publishing results on the platform and elsewhere. But then he felt the need to scale-up the work, and he realised it was not going to be possible in a journalism or advocacy set-up. Sutaria then committed fully to the startup, in the knowledge that the

> segment of people that we engage with requires the scientific validation of the realtime air quality data that we were sensing and analysing. And so, the majority of our engagement is with the scientific air quality and aerosol research community who have been investing in these technologies and analytics dashboards.
>
> (Personal communication, 2020)

We see with Sutaria a move towards airpreneurship through a deeper engagement with the technosciences. On the other hand, Mrutyunjay Mishra's views on activism were far stronger. While Mishra's interests hinged around using open-data for social and environmental change, he viewed change consequent to entrepreneurial rather than activist work. In Mishra's words,

> I am not an open data activist. I would probably call myself an open data evangelist. Not activist. I'm not an activist. I stay away from activists. It's very simple . . . activism is sometimes spilling blood all over the place. But I know that if I do sustainable businesses, and if I keep my ethics and values clean, I can do things far better and easier. I can work with everybody if I am the businessman.
>
> (Personal communication, 2017)

Mishra's point is that for maximum impact, ecopreneurs needed to do their core job well, rather than stray on to other fields of action. Interestingly, his personal trajectory was interspersed with contact with environmental activists at an early age through his father, who was a prominent activist in Odisha. Mishra perhaps felt less need to buffer his business interests with philanthropic or political ones.

Ecopreneurship and the public interest

Our objective in this chapter has been to parse out airpreneurship in Delhi without viewing these shifts through frameworks, whether celebratory or highly critical, received from elsewhere. We have shown the different arcs of involvement in the sphere as well as the airpreneurs' motivations and investments. Here, we take up the final question for the chapter, that is, the implications of these developments on public interest.

While there is no doubt that in isolation, the consumerism involved in the proliferation of anti-pollution commodities seems problematic, we offer two arguments towards complicating this reading. First, since the air *is* objectively dangerous, it is gratuitous to blame those who purchase these gadgets to protect themselves and their families. It is no surprise that residents use whatever is at hand and within means to lessen their risk, and they certainly should not be judged for doing so. Second, knowing the extent of pollution via handheld monitors or cocooning oneself indoors with purifiers does not imply an experience of luxury, as with other kinds of conspicuous consumption, but helplessness, even paralyses, given the disruption of the taken-for-granted everydayness of life. This might explain why experiencing toxic air through these gadgets seems to create a consciousness of vulnerability and the impulse to transcend individualism and enter into the public discourse on air pollution, whether through institutions, events, or the social media. Among the airpreneurs too, we find a desire to move beyond their individual business and build larger collaborations towards public interventions.

In their provocative essay on scientific citizenship, Elam and Bertilsson (2003) discuss the fraught engagement between the public and science in the time of "post-normal science," marked by uncertainty, contested values, and the simultaneous urgency of action (p236). Two models of this engagement are presented by the authors: the first, based on Schumpeterian ideas discussed earlier, is an innovative scientific entrepreneur who uses specific "practical talents and virtues" to bring change to technoscientific practice. It has been argued that much collective faith has been placed in the Indian context on the "visionary entrepreneur" (Chakravartty and Sarkar, 2013). Certainly, a trend that we have identified in the previous sections is some of Delhi's airpreneurs viewing themselves as change agents *qua* individual green evangelists. Kamal Meattle, for instance, told us about his run-ins with several influential people, from the prime minister to the US ambassador in India, during his personal environmental battles. In the long run, however, we view the key to progressive change and public interest in the creation and sustenance of multidisciplinary and multisectoral networks and collaborations. This is the second model of change, as per Elam and Bertilsson (2003), of a wider participation of society in shaping the meaning of virtue in decision-making processes, a kind of extended peer review (p236). The authors further note that "some of the most important individual and collective actors dedicating themselves to the task of 'contextualising' scientific knowledge in society, will not themselves originate from within the established bounds of science" (p237). It is here that we see the greatest possibilities of airpreneurship in Delhi, especially that emanating from the tech-data arc.

Few examples: Rohit Bansal's Purelogics Lab has installed monitors across several locations around Delhi and beyond, which contribute to generating a fine-grained picture of pollution, easily accessible via their phone app. Bansal is active on social media, putting out visualisations and generating debate amongst readers. Mishra's Indian Open Data Association banks on the free availability of environmental data to allow differently situated individuals *see* the problem from their unique vantage points and bring a diversity of views on to the table. Mishra says, "let's make the data public first. Once the data [are] public, then some creative [person] will think something and make it actionable" (personal communication, 2017). Ronak Sutaria's Respirer Living Sciences has partnered with the Indian Institute of Technology-Kanpur, and the industry-funded Shakti Sustainable Energy Foundation in 2017 to develop and deploy low-cost monitors in ten Indian cities. This initiative, to be viewed alongside Urban Emissions' ApNA project discussed in Chapters 3 and 4, is related to the NCAP rolled out by the Government of India, which aims to advance monitoring in over 100 cities across the country. In its over 2 years of operation, the project has generated scientific publications contributing to the global monitoring debate, and enabled conversations with diverse institutions from local governments and different SPCBs. The project initially involved testing and calibration of the newly produced monitors with reference-grade monitors to enhance the accuracy and robustness of the machines. Thereafter, data from the various study sites were used to develop a fine-grained temporal picture of air pollution, apprising policymakers of the situation in their respective locality, and handing information through apps and social media to concerned citizens for personal action and collective mobilisations. Similarly, Kaiterra has contributed its outdoor air quality monitors to researchers at the EPoD India at the Institute for Financial Management and Research, and the Energy Policy Institute at the University of Chicago (EPIC)'s Delhi unit to relay data for analysis, which allows "very dense reading, and develop very accurate heat maps of the air pollution levels of the city" (Nita Soans, personal communication, 2019). EPIC has, in turn, partnered with the Delhi Government to identify and incubate air quality interventions. The move in environmentalism towards data-drivenness, low-tech solutions, and open-source technologies has the potential to democratise the research-development process (Kera, 2011). While it is difficult to predict what the future may hold for the collaborations discussed here, they certainly encourage conversations on pollution from different vantage points and generate data-based oversight of state regulators.

To conclude, in this chapter, we have argued for the need to situate globally circulating concepts like ecopreneurship in their specific regional context. The actual practice of ecopreneurship taking shape in Delhi implies new subjectivities in both environmentalism and entrepreneurship. With ecological degradation, the market for ameliorating technologies has grown tremendously, and these commodities, therefore, become frontiers for entrepreneurial action. At the same time, with the rise of smart urbanism, environmentalists have to engage deeply with data and technoscientific artefacts like air quality monitors. It is at the confluence of these processes that the specificities of Delhi's airpreneurship are located. There are clear gaps in participation, since it requires resources and technical knowledge, but at the same time, questions of public interest are being thought through, and responses are being articulated and put into practice, making this a highly dynamic space of experimentation and collaboration.

Notes

1 Barun Agarwal, CEO of Breathe Easy Labs, quoted in Chhabra, 2015.
2 A private company can be contracted in Delhi to conduct air testing for home for INR 1,500 for a couple samples of PM2.5 and CO_2 and INR 10,000 for several samples of PM2.5, CO_2, PM10, HCHO, CO, and VOCs. With this knowledge, families can purchase appropriate purifiers to make air reasonably safe to breathe.
3 As reported in Economic Times, 15 December 2019. Available at https://eco nomictimes.indiatimes.com/industry/cons-products/durables/air-purifier-sales-see-up-to-60-spike-on-rising-air-pollution/articleshow/72671867.cms?from=mdr (Accessed 11 August 2020).
4 Birth of Airveda. Available at www.airveda.com/blog/birth-of-airveda (Accessed 21 August 2020).
5 Interview with Jai Dhar Gupta. Available at www.youtube.com/watch?v= ljPRcEpaL7o (Accessed 20 August 2020).
6 Graphic story. Available at www.aqi.in/about-us#aboutus (Accessed 12 May 2020).
7 Namita Gupta profile. Available at https://tecno.dailyhunt.in/news/india/english/ yourstory-epaper-yourstory/how+this+iit+delhi+alumna+is+fighting+pollution +with+air+quality+monitors-newsid-n178271530?pgs=N&pgn=4&&nsk=herst ory-updates-herstory (Accessed 21 August 2020).

References

Acharya, N. (2019) 'Today's social entrepreneur: inspired by Gandhi, taught by Prahalad, leading like Yunus', *Forbes*, 19 September. Available at www.forbes. com/sites/nishacharya/2019/09/19/todays-social-entrepreneur-inspired-by-gandhi-taught-by-prahalad-leading-like-yunus/?sh=30ebca5db8c7 (Accessed 2 July 2020).
Chakravartty, P. and Sarkar, S. (2013) 'Entrepreneurial justice: the new spirit of capitalism in emergent India', *Popular Communication*, 11(1), p58–75.

Cherry, M. A. and Sneirson, J. F. (2012) 'Chevron, greenwashing, and the myth of "green oil companies"', *Journal of Energy, Climate and the Environment*, 3, p133–154.

Chhabra, E. (2015) 'Selling fresh air in the world's most polluted city', *Next City*, 13 July. Available at https://nextcity.org/features/view/south-asia-india-air-pollution-clean-air-solutions-delhi (Accessed 11 March 2020).

Coutard, O. and Guy, S. (2007) 'STS and the city: politics and practices of hope', *Science, Technology, & Human Values*, 32(6), p713–734.

Datta, A. (2018) 'The digital turn in postcolonial urbanism: smart citizenship in the making of India's 100 smart cities', *Transactions of the Institute of British Geographers*, 43(3), p405–419.

de Bercegol, R. and Gowda, S. (2018) 'A new waste and energy nexus: rethinking the modernisation of waste services in Delhi', *Urban Studies*, 56(11), p2297–2314.

Demaria, F. and Schindler, S. (2016) 'Contesting urban metabolism: struggles over waste-to-energy in Delhi, India', *Antipode*, 48(2), p293–313.

Elam, M. and Bertilsson, M. (2003) 'Consuming, engaging and confronting science: the emerging dimensions of scientific citizenship', *European Journal of Social Theory*, 6(2), p233–251.

Farinelli, F., Bottini, M., Akkoyunlu, S. and Aerni, P. (2011) 'Green entrepreneurship: the missing link towards a greener economy', *ATDF Journal*, 8(3–4), p42–48.

Follman, A. (2016) 'The role of environmental activists in governing riverscapes: the case of the Yamuna in Delhi, India', *South Asia Multidisciplinary Academic Journal*, 14. Available at https://journals.openedition.org/samaj/4184 (Accessed 28 June 2020).

Fortun, K. (2009) *Advocacy after Bhopal: environmentalism, disaster, new global orders*. Chicago: University of Chicago Press.

Foster, J. B. (2000) *Marx's ecology: materialism and nature*. New York: New York University Press.

Gill, K. (2009) *Of poverty and plastic: scavenging and scrap trading entrepreneurs in India's urban informal economy*. New Delhi: Oxford University Press.

Goldstein, J. (2018) *Planetary improvement: cleantech entrepreneurship and the contradictions of green capitalism*. Cambridge, MA: The MIT Press.

Harvey, D. (1989) 'From managerialism to entrepreneurialism: the transformation in urban governance in late capitalism', *Geografiska Annaler: Series B*, 71(1), 3–17.

Haugh, H. (2005) 'A research agenda for social entrepreneurship', *Social Enterprise Journal*. Available at www.emerald.com/insight/content/doi/10.1108/17508610580000703/full/html (Accessed 30 July 2019).

Hughes, S. (2013) 'Justice in urban climate change adaptation: criteria and application to Delhi', *Ecology and Society*, 18(4). Available at www.ecologyandsociety.org/vol18/iss4/art48/ (Accessed 1 April 2020).

Irani, L. (2019) *Chasing innovation: making entrepreneurial citizens in modern India*. Princeton, NJ: Princeton University Press.

Kera, D. (2011) 'Grassroots R&D, prototype cultures and DIY innovation: global flows of data, kits and protocols', *Pervasive Adaptation*, 51. Available at https://citeseerx.ist.psu.edu/viewdoc/download?doi=10.1.1.446.8306&rep=rep1&type=pdf#page=51 (Accessed 11 June 2020).

Kimura, A. H. (2016) *Radiation brain moms and citizen scientists: the gender politics of food contamination after Fukushima*. Durham, NC: Duke University Press.

Levenda, A. M. and Tretter, E. (2019) 'The environmentalization of urban entrepreneurialism: from technopolis to start-up city', *Environment and Planning A*. Available at https://doi.org/10.1177/0308518X19889970 (Accessed 25 April 2020).

Luke, T. W. (1998) 'The (un) wise (ab) use of nature: environmentalism as globalized consumerism', *Alternatives*, 23(2), p175–212.

Mertia, S. (2017) 'Socio-technical imaginaries of a data-driven city: ethnographic vignettes from Delhi', *The Fibreculture Journal*. Available at http://fibreculture journal.org/wp-content/pdfs/FCJ-217SandeepMertia.pdf (Accessed 29 March 2020).

Moore, J. W. (2015) *Capitalism in the web of life: ecology and the accumulation of capital*. New York: Verso Books.

O'Neill, K. and Gibbs, D. (2016) 'Rethinking green entrepreneurship: fluid narratives of the green economy', *Environment and Planning A: Economy and Space*, 48(9). Available at https://journals.sagepub.com/doi/10.1177/0308518X16650453 (Accessed 12 June 2020).

Perng, S. Y., Kitchin, R. and Mac Donncha, D. (2018) 'Hackathons, entrepreneurial life and the making of smart cities', *Geoforum*, 97, p189–197.

Phillips, M. (2012) 'On being green and being enterprising: narrative and the ecopreneurial self', *Organization*, 20(6), p794–817.

Sadowski, J. (2020) 'Who owns the future city? Phases of technological urbanism and shifts in sovereignty', *Urban Studies*. Available at https://doi.org/10.1177/0042098020913427 (Accessed 22 August 2020).

Schaper, M. (2002) 'The essence of ecopreneurship', *Greener Management International*, 38, p26–30.

Sivaramakrishnan, K. (2017) 'Courts, public cultures of legality, and urban ecological imagination in Delhi', in Rademacher, A. and Sivaramakrishnan, K. (eds.) *Ecologies of urbanism in India: metropolitan civility and sustainability*. Hong Kong: Hong Kong University Press, p137–161.

Śledzik, K. (2013) 'Schumpeter's view on innovation and entrepreneurship', in Hittmar, S. (ed.) *Management trends in theory and practice*. Zilina, Poland: University of Zilina.

Sundar Rajan, K. (2006) *Biocapital: the constitution of postgenomic life*. Durham, NC: Duke University Press.

Thompson, N., Kiefer, K. and York, J. G. (2011) 'Distinctions not dichotomies: exploring social, sustainable, and environmental entrepreneurship', in Lumpkin, G. T. and Katz, J. A. (eds.) *Social and sustainable entrepreneurship*. Bingley: Emerald, p201–229.

6 Postscript

COVID-19, air pollution, and environmentalism

We wrote most of this book at a strange moment in time. The COVID-19 pandemic emerged as suddenly as it disrupted economies and lives around the world. It hit India particularly hard. Towards the end of March 2020, learning from the experiences of China and Europe, the Indian Government enforced a nationwide lockdown, practically halting economic activity and the movement of people. A media blitz highlighting the dangers of the novel coronavirus and the responsibility of citizens to remain indoors and protected was mounted alongside. With the exception of essential services and large industries, almost all economic activity ground to a sudden halt, which lasted for the next 2 months. A series of "unlock" policies later progressively reopened the economy, though some activities remained shut, over 6 months after the first lockdown. As we finish the book, Delhi is witnessing record number of COVID-19 cases, which have been attributed to the rising air pollution as the winter sets in (Ethiraj, 2020).

Apart from all else, the lockdown effectively cleaned the nation's air (Mahato et al., 2020), though pollution is back as the winter sets in and economic activities restart. The clean air during the lockdown attracted frenzied interest from researchers across disciplines. Scientific work at the interface of COVID-19 and air has been of primarily three types: (1) on the transmission of the virus via aerosols; (2) on the impact of poor air quality on health outcomes due to COVID-19; and (3) on the impact of economic shutdowns on air quality. After the initial belief that coronavirus transmission took place only via droplets, evidence generated by aerosol experts suggested that it also spread through aerosols, which made it highly infectious. On the second trajectory of research, COVID-19 death rates have been shown to be significantly correlated with the intensity of air pollution (Ching and Kajino, 2020). In an Indian study, Chakraborty et al. (2020) found that chronic exposure to NO_2 caused damage to the lungs and may play "a major role in prognosis of the patient once infected," adding that "the homeless, poverty-stricken Indians, hawkers, roadside vendors, and many others who are regularly exposed to vehicular exhaust, may be at a higher risk in the COVID-19 pandemic."

More directly related to our concern, the lockdown led to an intense conversation on the reduced pollution observed around the world. In the words of Sarath Guttikunda, whom we profiled in Chapter 3,

> From a research viewpoint, this is a fantastic experiment . . . The big cities always point outwards, saying, 'All my pollution is coming from outside. Now we don't have to blindly say, 'Look, you are responsible for seventy percent of your pollution. Please do something about it. We have that proof.
>
> (Quoted in Karnad, 2020)

In their analysis of air quality data before and during the lockdown across several Indian cities, Guttikunda and Nishadh (2020) found high degree of reduction in PM2.5, PM10, and NO_2, 10–60 per cent *increase* in daytime ozone concentrations, and almost no difference in SO_2 concentrations. The reasons are logical. The first three pollutants are released into the atmosphere by vehicles, construction, and industry—largely shut down during the lockdowns—while SO_2 pollution is caused primarily by power plants, mining, and steel industries, which, by and large, continuously operated during this time. Ozone concentration was interestingly higher due to the warming weather and also its chemistry, since lower concentration of NO leads to reduced ozone titration. Similar findings were reported for European and Chinese cities (Sicard et al., 2020; Venter et al., 2020). It should be added here that even as ambient air pollution largely declined, indoor pollution actually worsened, in part, because people looked to save money by opting for biomass-based fuels in place of LPG (Tripathi, 2020).

Still, images of blue skies over Delhi and other cities went viral alongside photographs of the Himalayas seen clearly from faraway places like Jalandhar and Saharanpur. Such visuals have been interpreted as a "blessing in disguise" (Gautam, 2020; Ghosh and Ghosh, 2020) and a "kiss of life" (Karnad, 2020) for Indian cities. International publications like the BBC, CNN, and New York Times too ran stories on how Delhi was enjoying a rare experience of clean air because of the lockdown. "It's positively Alpine!" screamed a Guardian headline. Stepping back from this giddy debate on COVID-19 and air pollution, a set of questions emerge at this juncture, which we take up in this postscript: How should environmentalists approach the moment? And what of the gains in advocacy that have been painstakingly achieved, given the pandemic-induced changes to economic priorities?

We believe overall that this is a productive moment for environmental politics with the public eye trained on the atmosphere. For one, scientific studies conducted during this time have decisively shown that while meteorology and geography are important to air quality, the bulk of the pollution is caused by socially produced sources, chiefly, vehicles, industry, and construction. According to Guttikunda and Nishadh (2020), one of the lessons of the lockdown's air is that it has "demonstrated to the public that the only way to achieve 'clean air' is by cutting emissions at the source." This point has been forcefully made by several advocates in their criticism of the Supreme Court's insistence on the installation of smog towers in Delhi. They have argued that removing toxic elements once they are already in the atmosphere over a large region is next to impossible, apart from being a highly expensive enterprise. As Santosh Harish of CPR argues (2020),

> the use of public funds becomes more efficacious and sustainable when we implement permanent solutions like improving public transport and supporting related activities, making waste segregation, collection and management more effective, and promoting the use of cleaner fuels like LPG for cooking and heating.

Second, the cleaner air invites urban residents to visualise the possibilities of a healthier future, which not only laypeople but also advocates have found extremely tough to imagine and communicate. This imaginary was powerfully articulated by an activist organisation to campaign for "Blue Skies Beyond the Lockdown: Saal Bhar 60 [60 AQI all year round]." The 12-year-old environmental activist, Ridhima Pandey, who has been a part of this campaign, believes that

> in the past two months, it has felt like all the things my generation has been fighting were gifted to us—blue skies, low emissions, clean air. This only means that an AQI of Saal Bhar 60 is possible. Our government needs to treat it like the emergency it is, and have strict timelines for bringing down pollution levels.
>
> (Behl, 2020)

The discourse of possibilities has, in this way, opened the space for a conversation on air pollution that pushes for goals beyond the very modest and muted ones that activists and advocates have grown accustomed to, given the state's developmentalist bent and economic priorities.

In a critical piece, anthropologist Nayanika Mathur (2020) criticises the genre of environmental writing described earlier, that is, the "renewal of nature during COVID-19," since clean air in cities like Delhi has been concomitant with massive misery for the urban majority. Mathur writes: "There is a direct line of causality between the bluer sky that we are seeing—caused by the sudden and intense lockdown—and the deaths and mass suffering unleashed by a hastily imposed, ill-planned, and ultimately failed lockdown." Our contention is that rather than placing these at opposing ends of urban citizenship, we must call for simultaneous attention to *both* the blue skies and the livelihood disruptions and the general hardship. As we describe throughout the book, air pollution advocacy in the Delhi region is thinking with these questions, and there are signs that environmental justice is being seen as a vital element of the emergent atmospheric citizenship via translations, collaborations, and an overall emphasis on greater inclusiveness.

Having said so, there are a few pressing challenges to progressive environmental politics at this moment. Advocates have to find ways to ensure that longer struggles to promote public transportation and other healthier modes of mobility remain important to revisions in urban policy brought about by the coronavirus. During the lockdown, according to data shared by CSE, while there was a drop of 60 to 80 per cent in the total number of walking and driving trips in selected Indian cities, and in contrast to the pre-lockdown trends, walking trips exceeded driving trips through April 2020.[1] The situation would certainly have changed as public transportation reopened, but the fear of infection from the virus in crowded buses and metro systems may push people towards individualised transport like cars and two-wheelers. Added to this is the state's technomanagerial push towards electric vehicles (EVs), with far less consideration of other less expensive, public, and relatively cleaner measures. Since power generation is likely to continue to be heavily dependent on coal, any push towards EVs is only going to exacerbate ecological damage and air pollution in the country's mining belt, while further reinforcing the disproportionate emphasis on metros in the pollution debate. Cities like London and Paris have pushed for greater space and infrastructures for cyclists in the post-lockdown city, but in India, all the recent initiatives to promote cycling—largely associated with Smart City initiatives—are directed at recreational cycling by wealthier residents with hardly any attention to those who cycle on an everyday basis, because they have no other alternative, given the rising costs of public transport. Accounting for purchasing power, the Delhi Metro, for instance, is now more unaffordable than the New York Subway (Somvanshi, 2018). The Delhi Government's decision to make buses free of cost for women is certainly a step in the right direction and encourages a relatively safer and greener mode of transportation than many would otherwise have opted for. Attention is also due to modes like the Grameen Sewa, so critical to the urban majority, but under the radar of policy.

As the country and cities open for business once again, given the tremendous losses incurred due to the lockdown, there is pressure to take the easiest means towards an economic recovery, even if it means compromising on environmental regulations and long-term climate change adaptation and mitigation measures (McNeill, 2020). Energy transition to renewables is likely to be stalled in favour of continued use of the cheapest forms of power production, which inevitably leads to coal. As of September 2020, a vigorous debate on the revised Environmental Impact Assessment draft legislation is also underway, with advocates calling out the loosening of regulations and its openly 'business-friendly' bias. For instance, "the proposed notification allows the government to grant an ex post facto environmental clearance to projects that have commenced illegally without a clearance" (Ghosh, 2020). The proposed changes also aim to lessen the importance assigned to the public consultation process in the existing law. These shifts must be understood as part of a larger onslaught directed against laws that sought to protect labour and the environment. As the lockdowns were lifted, many states hurriedly made changes to extant labour laws, to further weaken already precarious protections for documented and informal workers in urban areas (Mander and Verma, 2020). The environment and labour must, therefore, be considered as part of a conjoined struggle not only to stand up against the anticipated post-lockdown revanchist state policy, but also to work towards an urban environmentalism that sutures the divides which have characterised it until very recently.

Slowness and carefulness as collaborative tactics

Our book has shown how this is possible through a prolonged engagement with air in Delhi by bringing out the dilemmas that experts, advocates, governments, and businesses face while grappling with a slowly unfolding disaster. As the conversation moves from Delhi to smaller towns and rural communities (Chapter 3), questions of social justice find greater prominence. But we do not need to distance from the capital region to talk about social justice. Environmental advocacy in Delhi can no longer be in silos of environmentalism and social justice. The pandemic makes that even more obvious. As the conversation on air moves between different scales for governance, we must watch out for what is being offered and to whom. In Chapter 4, we write about the challenges of coordinational capacity when it comes to governing air, which becomes more pertinent in this moment. Chapter 5 considers how people try to protect themselves from toxic air with air purifiers and masks, and ecopreneurs enter the market with the intention of making profit and doing social service. We explore their backstories to understand their motives and how they want to reconcile profit-making with the larger public interest. As people live increasingly insular lives in the pandemic shielded by protective gear, already fragmented breathing spaces will become even more fragile. They will try to innovate new ways of maintaining, fixing, and doing, which will not be perfect and easy. We must attend to how people slowly try to build, maintain, and collaborate as they face a prolonged disaster. These efforts could fall apart and be predictably prone to social critique. But through our interlocutors and colleagues, we have come to understand that slow advocacy and collaboration are responsive to failure of systems, because they build on those failures. The way forward is not to figure out a perfect advocacy style or seamless collaboration. It is to evolve tactics where uncertainties, fumbling, and maintenance are as respected as leaping ahead.

To that end, we propose two tactics here—slowness and carefulness. Slowness, as a tactic, is not just about doing things incrementally over a long time. It is about muddling through and figuring out where the limits of our understanding lie. It means inviting into conversation people whose thought styles might interrupt our styles of thinking and doing. Slowness also acknowledges the slow violence of toxicity. The concept of slow violence was proposed by Rob Nixon (2011) to ask environmentalists to consider how they frame the future of their advocacy. Do they advocate for a shared planetary future where all humans are equal and have equal responsibility? Does apocalyptic thinking about the future foreclose engagements on the ground? When environmental advocates talk about the future their children and grandchildren would inherit, what about people who cannot and will not reproduce? When they talk about mortality and life expectancy and inadvertently value an able-bodied human, what about those with debility? Nixon's work documents the legacies of environmentalism in the Global South that environmental advocates may build on.

Monamie Bhadra (2013) also documents strands of activism that our interlocutors have not yet engaged sufficiently with. The Jharkhandi Organisation Against Radiation (JOAR) has been working for 30 years to seek environmental justice for indigenous Santhal and Ho communities, who live and work near uranium mines, resisting the Department of Atomic Energy's denials of wrongdoing. Bhadra writes about the activists-scholars who have organised around anti-nuclear resistance, crafting their advocacy in terms of land grabs and livelihoods. Abraham and Rajadhyaksha (2015) mention the villagers of Baliapal in Orissa who blocked access to roads through warning systems for years to resist building a new missile testing range. Our attempt here is not to point out that these styles of activism are more virtuous, but that they matter in the locations environmental advocates find themselves interested in.

Slowness could also be used by environmental advocates to frame questions around data differently. At present, environmental advocates often assume that they know what to communicate, what to find out, and what to figure out when it comes to science for public understanding. This tends to create platforms and databases that reflect prominent trends towards capturing and visualising particulate matter. In doing so, they work within the given global conditions of science and technology *whereas* they could use their expertise and advocacy to connect quantitative data to sensorial and ethnographic data to make visible the people they advocate for in all their humanness and complexity. They could also search for already-existing capacities for knowing, understanding, and archiving data for anthropocenic futures (Schütz, 2019).

Our second tactic of carefulness comes from feminist geographers and STS scholars (Martin et al., 2015; Mattern, 2018) who engage with practices of care while also asking who cares and why. Advocacy at the core is a matter of care and maintenance, of slowly building collaborations through careful attention and practice. In 2020, one of us (Prerna) was invited by Pallavi Pant to join the transnational group Women in Air Quality in South Asia (WAQSA). Initiated informally through conversations between interlocutors who are now friends and acquaintances, WAQSA's mandate is to promote and support women working in South Asian air quality. The diversity among Indian scientists is also now being seriously examined (Thomas, 2020). Environmental advocates including us are usually upper-caste, and a slow violence done by upper-castes has been to construct themselves as casteless beings. This is especially relevant for studies of air pollution advocacy, because who is seen as polluting and polluted is also a matter of caste.

Care is a question of power. To be careful is to recognise one's own privilege and limitations of understanding. But it is also to acknowledge that when doing advocacy, one is advocating for one's own irrelevance and vacating space for something and someone else. Just as environmental advocates build alliances with social and environmental justice movements, they should also be ready to leave space for the experiences and locations that remain unarticulated in science and policy.

Note

1 This was shared during a webinar entitled "Reinventing Public Transport and Mobility in the New Normal," 25 May 2020.

References

Abraham, I. and Rajadhyaksha, A. (2015) 'State power and technological citizenship in India: from the postcolonial to the digital age', *East Asian Science, Technology and Society*, 9(1), p65–85.

Behl, M. (2020) '12-year-old climate activist launches clean air campaign', *Times of India*, 24 May. Available at https://timesofindia.indiatimes.com/city/

nagpur/12-year-old-climate-activist-launches-clean-air-campaign/article-show/75927886.cms (Accessed 19 July 2020).

Bhadra, M. (2013) 'Fighting nuclear energy, fighting for India's democracy', *Science as Culture*, 22(2), p238–246.

Chakraborty, P. et al. (2020) 'Exposure to nitrogen dioxide (NO2) from vehicular emission could increase the COVID-19 pandemic fatality in India: a perspective', *Bulletin of Environmental Contamination and Toxicology*, 105, p198–204.

Ching, J. and Kajino, M. (2020) 'Rethinking air quality and climate change after COVID-19', *International Journal of Environmental Research and Public Health*, 17(14), p5167.

Ethiraj, G. (2020) 'Coronavirus: the record number of new cases in Delhi is a "direct effect of air pollution"', *Scroll.in*, 14 November. Available at https://scroll.in/article/978454/coronavirus-the-record-high-number-of-new-cases-in-delhi-is-a-direct-effect-of-air-pollution (Accessed 14 November 2020).

Gautam, S. (2020) 'The influence of COVID-19 on air quality in India: a boom or inutile', *Bulletin of Environmental Contamination and Toxicology*, 104, p724–726.

Ghosh, S. (2020) 'Mandate betrayed: draft EIA notification dilutes environmental protections, is in denial of ecological crises', *Indian Express*, 11 August.

Ghosh, S. and Ghosh, S. (2020) 'Air quality during COVID-19 lockdown: blessing in disguise', *Indian Journal of Biochemistry and Biophysics*, 57, p420–430.

Guttikunda, S. K. and Nishadh, K. A. (2020) *Air quality trends and lessons learnt in India during the COVID lockdowns*. Policy Brief, CCAPC/2020/03. New Delhi: Collaborative Clean Air Policy Centre.

Harish, S. (2020) 'Smog towers have no scientific basis as a policy measure', *The Wire*, 6 August. Available at https://science.thewire.in/environment/delhi-smog-tower-iitb-neeri-evidence-judiciary-policy/ (Accessed 30 September 2020).

Karnad, R. (2020) 'The coronavirus offers a radical new vision for India's cities', *New Yorker*, 13 April. Available at www.newyorker.com/news/dispatch/the-coronavirus-offers-a-radical-new-vision-for-indias-cities-pollution (Accessed 21 August 2020).

Mahato, S., Pal, S. and Ghosh, K. G. (2020) 'Effect of lockdown amid COVID-19 pandemic on air quality of the megacity Delhi, India', *Science of the Total Environment*, 730, p1–23.

Mander, H. and Verma, A. (2020) 'The coronavirus lockdown has been a war on India's informal labour', *The Wire*, 22 August. Available at https://thewire.in/labour/coronavirus-lockdown-informal-labour (Accessed 27 August 2020).

Martin, A., Myers, N. and Viseu, A. (2015) 'The politics of care in technoscience', *Social Studies of Science*, 45(5), p625–641.

Mathur, N. (2020) '"Nature is healing": why we need to be careful about how we tell the story of the pandemic', *Scroll.in*, 6 June. Available at https://scroll.in/article/963743/nature-is-healing-why-we-need-to-be-careful-about-how-we-tell-the-story-of-the-pandemic (Accessed 7 July 2020).

Mattern, S. (2018) 'Maintenance and care', *Places Journal*. Available at https://placesjournal.org/article/maintenance-and-care?cn-reloaded=1 (Accessed 5 September 2020).

McNeill, V. F. (2020) 'COVID-19 and the air we breathe', *ACS Earth Space Chemistry*. Available at https://doi.org/10.1021/acsearthspacechem.0c00093 (Accessed 20 August 2020).

Nixon, R. (2011) *Slow violence and the environmentalism of the poor*. Cambridge, MA: Harvard University Press.

Schütz, T. (2019) 'Archiving for the anthropocene: notes from the field campus', *Platypus: The CASTAC Blog*. Available at http://blog.castac.org/2019/10/archiving-for-the-anthropocene-notes-from-the-field-campus/ (Accessed 18 August 2020).

Sicard, P. et al. (2020) 'Amplified ozone pollution in cities during the COVID-19 lockdown', *Science of the Total Environment*, 735. Available at https://doi.org/10.1016/j.scitotenv.2020.139542 (Accessed 12 September 2020).

Somvanshi, A. (2018) 'Yes, Delhi metro is more unaffordable than the New York city subway', *Down to Earth*, 24 September. Available at www.downtoearth.org.in/blog/air/yes-delhi-metro-is-more-unaffordable-than-the-new-york-city-subway-61705 (Accessed 2 August 2020).

Thomas, R. (2020) 'Brahmins as scientists and science as Brahmins' calling: caste in an Indian scientific research institute', *Public Understanding of Science*, 29(3), 306–318.

Tripathi, B. (2020) 'Coronavirus: while Indians were rejoicing over clear blue skies, indoor pollution was rising', *Scroll.in*, 12 June. Available at https://scroll.in/article/967106/coronavirus-while-indians-were-rejoicing-over-clear-blue-skies-indoor-pollution-was-rising (Accessed 18 August 2020).

Venter, Z. S. et al. (2020) 'COVID-19 lockdowns cause global air pollution declines', *Proceedings of the National Academy of Sciences of the United States of America*, 117(32), p18984–18990.

Index